W9-AVB-469

Frederick the Great

INSTRUCTIONS FOR HIS GENERALS

Translated by
General Thomas R. Phillips

DOVER PUBLICATIONS, INC.
Mineola, New York

Bibliographical Note

This Dover edition, first published in 2005, is an unabridged republication of the work originally published by The Military Service Publishing Company, Harrisburg, Pennsylvania, in 1944.

Library of Congress Cataloging-in-Publication Data

Frederick II, King of Prussia, 1712-1786.
[Des Königs von Preussen Majestät Unterricht von der Kriegs-Kunst an seine Generals. English]
Instructions for his generals / Frederick the Great ; translated by General Thomas R. Phillips.
p. cm.
Originally published: Harrisburg, Pa. : Military Service Pub. Co., 1944.
ISBN 0-486-44403-1 (pbk.)
1. Generals—Germany—Prussia—Handbooks, manuals, etc. 2. Prussia (Kingdom). Armee—Officers' handbooks. 3. Military art and science—Handbooks, manuals, etc. 4. Command of troops—Handbooks, manuals, etc. I. Title.

UB200.F74314 2005
355.02—dc22

2005045465

Manufactured in the United States of America
Dover Publications, Inc., 31 East 2nd Street, Mineola, N.Y. 11501

Contents

FREDERICK THE GREAT; 1712-86.
Painted by Kaulbach

Introduction

FREDERICK the Great, King of Prussia, was the founder of modern Germany. When he became King of Prussia it was a small state, in two parts, and of minor importance among the great powers of Europe. During his reign the population increased from 2,240,000 to 6,000,000, and the territory was increased by nearly thirty thousand square miles.

He fought all the great powers of Europe in the Seven Years' War and successfully defended the national territory.

Although he gained no new lands during this war his success against the overwhelming coalition of his enemies made him a great man of his time and the soldier whom all other soldiers in Europe were to imitate. Frederick's upbuilding of Prussia determined whether the many small German states eventually were to group themselves around Prussia or around Austria. The expansion of Prussia was continued by Bismarck.

Frederick the Great, or Frederick II, was the grandson of Frederick I, the first King of Prussia. His family The Hohenzollerns of Brandenburg, was founded by the Margrave Frederick, who began to rule in Brandenburg in 1415.

Successively the monarchs of the line and the years of their reigns and lives were:

1701-13, Frederick I; 1657-1713.

1713-40, Frederick William I, son of the above; 1688-1740.

1740-86, Frederick II, the Great; son of the above; 1712-1786.

1786-97, Frederick William II, nephew of the above; 1744-1797.

1797-1840, Frederick William III, son of above; 1770-1840.

1840-61, Frederick William IV, son of above; 1795-1861.

1861-1888, William I, brother of above; 1797-1888.

1888, March-June, Frederick III, son of above; 1831-1888.

1888-1918, William II, brother of above; 1859-1941.

Judgments of Frederick's qualities and merits, aside from those which bear upon his firmly-based renown as a military genius, differ widely amongst historians. The range of opinion runs from the extreme provided by Carlyle's unstinted praise and admiration, which he found it impossible to express adequately in less than the seven volumes of his *Frederick the Great,* and which consumed thirteen years of his study and time, to that of another critic who roundly condemned the Prussian monarch as "a canonized scoundrel."

Washington, Franklin and other American patriots of the Revolutionary era openly admired him. When he died Jefferson commented upon his death as an "European disaster" and as an event that "affected the whole world." Washington welcomed Baron Steuben, and his assistance as drillmaster and tactician, as one of Frederick's soldiers.

These Americans had reason to feel kindly toward Frederick on score of the service which he undoubtedly rendered the rebellious colonies during the dark winter of Valley Forge, 1777-8. German soldiers from Hesse and Ansbach had been hired by the British for service in America. Frederick who was outspoken against Germans being sent from their native states to fight across seas, refused permission for a considerable body of German mercenaries to pass through Prussian territory to ports of embarkation. Troops that left Anspach in September, 1777, did not reach New York until October, 1778.

Frederick Hampered Howe

To quote from Frederick Kapp's *Frederick the Great and the United States,* Leipsic, 1871: "Washington was suffering all the hardships of his winter quarters at Valley Forge from December, 1777 to June, 1778. His weak force could not withstand a vigorous attack by Howe, but when Howe learned of Frederick's prohibition of the passage of troops through Prussian territory he knew that that meant cutting off the prospects of any reinforcements. It was not the few men delayed in their journey that hampered Howe as much as the uncertainty about the coming of future German reinforcements. Frederick's policy was worth to Washington as much as an alliance, for it gave him time and helped to change the fortunes of war."

But, as Kapp explains, here "without really wishing to do so, Frederick rendered a real service to the young republic." Frederick was playing a double game. For the time, he was at odds with England. But later, when he needed England's help against Austria in the War of the Bavarian Succession (1778-9) he did authorize George the Third's hired soldiers to proceed to America, by way of Prussia. He refused to recognize, in advance of the complete triumph of the colonists, the government of the rebels or to receive officially any of its representatives.

After the Revolution the popularity in America of Frederick was attested by number of inns with signs bearing the legend "The King of Prussia." In Puritan New England and among the Germans of Pennsylvania and New York he was highly regarded as leader of Protestant resistance to Catholic aggression. Evidence is that Frederick never thought of anything but the interest of Prussia in the struggle between England and the American colonies.

One exploded legend is that, in token of his respect and admiration for Washington, he sent the American commander-in-chief a sword fashioned by a Prussian armorer. What happened was that a presentation sword was made by a German who commissioned his son to deliver it to Washington. The son left it in pawn to a Philadelphia tavern for thirty dollars. Friends of Washington redeemed it and it finally reached the General's hands.

Frederick's father, Frederick William, brought him up with extreme rigor in the hope that he would become a hardy soldier and "acquire thrift and frugality." To his father's disgust he showed no interest in military affairs and devoted his time to literature and music. He was so harshly treated by his father that he resolved to escape to England and take refuge there. He was helped by two friends, Lieutenant Katte and Lieutenant Keith. The plan was discovered, and the Crown Prince was arrested, deprived of his rank, tried by court-martial and imprisoned in the fortress of Kustrin. Keith escaped, but Katte was captured, tried and sentenced to life imprisonment. The King changed the

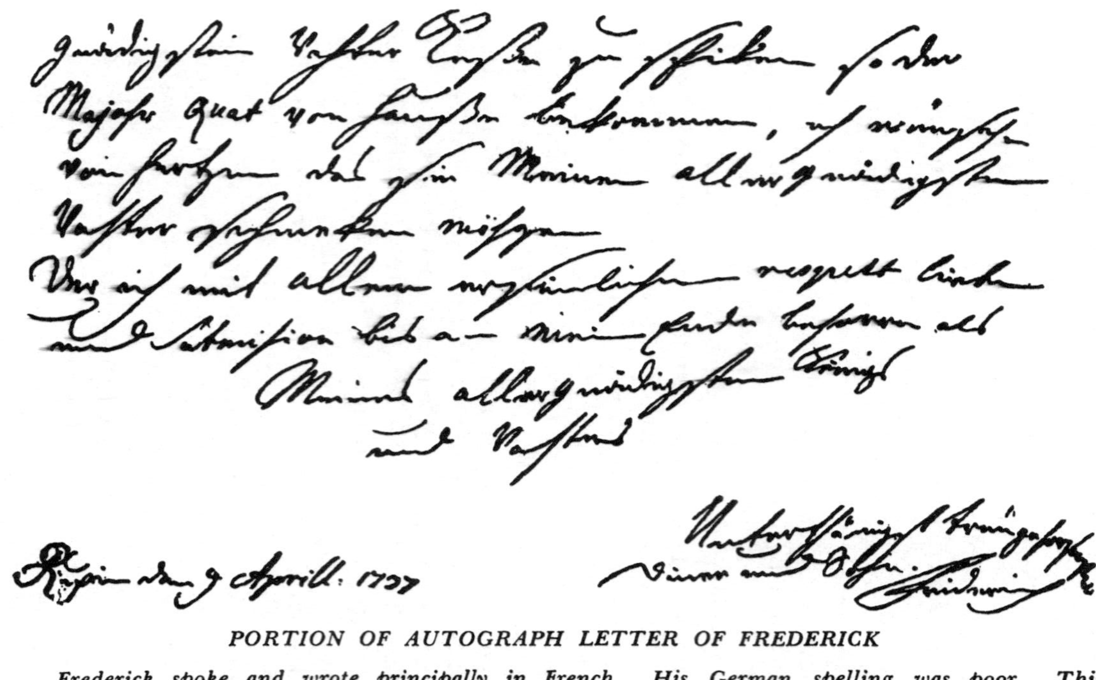

PORTION OF AUTOGRAPH LETTER OF FREDERICK

Frederick spoke and wrote principally in French. His German spelling was poor. This letter was written when he was twenty-five, and three years before he became King. It apparently was to, or intended for, his father, and to be transmitted to him through a Major Quat, a member of the King's household. Freely translated it reads: "To send greetings to my most gracious father, so that Major Quat may get them. I wish from my heart that they will be agreeable to my most gracious father, whom I love with all possible respect, and to the end of my life. My most gracious father and King, your most respectful, faithful servant and son, Frederick. Rupin, April 9th, 1737."

sentence and Frederick was forced to watch his friend beheaded.

Early Victories

While still restricted he was put to work in the auditing office of the war department checking invoices, payrolls, etc. He was allowed to appear in uniform a year later. He became King May 31, 1740, on the death of his father. He seized Silesia from Austria and gained a victory at Mollwitz, April 10, 1741. At this battle he fled from the field under the impression that it had been lost as a result of a furious charge of Austrian cavalry— a mistake which gave rise to a reputation that he lacked of personal courage.

He gained a second victory at Caslau on May 17, 1742, the war ending in the peace of Breslau and the cession of Silesia by Austria. He captured Prague in 1744, but was forced to retreat. In 1745 he won a series of victories and concluded the peace of Dresden, which for a second time assured him possession of Silesia.

Between this time and the commencement of the Seven Years War in 1756, he devoted himself to building up his kingdom. He restored the Academy of Sciences, encouraged agriculture, extended the canal system, drained and dyked the marshes of Oderbruch, stimulated manufactures and increased his army to 160,000 men. He worked arduously all his life, rising habitually at four or five o'clock in the morning and busying himself until midnight. He was economical in the conduct of his personal and state affairs and was scorned and lampooned for this. His economy was not understood in that age of glittering and useless royalty.

The Seven Years' War ended with the *status quo* established, but it was, in effect, a great victory. From this time on he devoted his energies to reconstitution of the devastated kingdom. Taxes were remitted for a time to the provinces that had suffered most. Army horses were distributed among the farmers and treasury funds were used to rebuild devastated cities. When he died August 17, 1786, he left an army of 200,000 men and 70,000,000 thalers in the treasury.

Frederick for his time, was a liberal, allowing freedom of press and speech and permitting publication of books and cartoons that derided and lampooned him. Riding along the Jäger Strasse one day he saw a crowd. "See what it is," he said to his groom. "They have something posted up about Your Majesty," said the groom, returning. Riding forward, Frederick saw a caricature of himself: "The King in very melancholy guise," (Preuss as translated by Carlyle), "seated on a stool, a coffee mill between his knees, diligently grinding with one hand and with the other picking up any bean that might have fallen. 'Hang it lower,' said the King, 'lower that they may not hurt their necks about it.'" The crowd cheered him as he rode slowly away.

He was deeply interested in the administration of justice and called himself "the advocate of the poor." It has been a fashion among recent writers to judge Frederick the Great harshly on the grounds of his diplomatic duplicity and the severity of his punishments in the army. But there is something to be said on the other side of that situation. He successfully played the game of deception common to all monarchs in his time. Like all the rest of them, he was intent on adding to his domain. Punishment was severe in the Prussian army, but so was it in all other armies. In the French army pillaging and desertion were punishable by death, whereas in the Prussian army flogging was the prescribed penalty for these crimes. Flogging was given up in the French revolutionary army, but it continued longer in the English army than it did in Germany. It was not until 1813 that flogging was abolished in the American army.

Saxe Gave Guidance

Saxe*, the French marshal, gained his successes in battles of position; he did not live long enough to take advantage of the mobility which his cadence gave to armies. Frederick was to profit by use of the new mobility of masses as Napoleon did later. He was the peer in this respect before

* See Saxe's *Reveries on the Art of War*, The Military Service Publishing Company.

Napoleon. He regularly marched his armies at a rate of twenty kilometers a day during periods of two, three and four weeks.

Armies of his time were more numerous, but not so well coordinated as formerly. Increases in numbers attenuated the personal tie between men and their commanders. Lines were thinner and fronts were extended, reducing control. This was the state in which Frederick found European tactics. At the same time his army was more perfect in details, thanks to the work of his father, and he spent more time on strategy.

Prussian soldiers, through constant drill, were able to load and fire their muskets twice as rapidly as their opponents. At the same time, they charged much more rapidly than the enemy and thus gained a further advantage in avoiding the enemy's fire. Until his enemies learned from him, these advantages helped greatly in his victories.

Frederick's tactics were based on mobility. This was especially favorable to him because the heavy, slow-moving battalions of his enemies lacked it. In his early years Frederick agreed with Saxe in his belief on the uselessness of fire in attack. Initially, and renewed from year to year, he ordered the soldiers to carry their guns on their shoulders in the attack and to fire as little as possible. But in 1758, he commenced to write that "to attack the enemy without procuring oneself the advantage of superior, or at least equal fire, is to fight against an armed troop with clubs, and this is impossible." Ten years later, in his *Military Testament,* he wrote: "Battles are won by superiority of fire." A decisive word, remarks Captain J. Colin, which marks a new era of combat.

Frederick was a sincere admirer of Saxe in his lifetime and his true testamentary executor. His writings advocate only the offensive, always, in every situation, in the whole of the operations as well as on the field of battle, even in the presence of a superior enemy.

"With such ardor," writes Capt. Colin, "how does he operate when he makes contact with the adversary? Not an immediate battle, but a long, drawn out jockeying for position. This was a characteristic of operations from ancient times until Frederick. Far from the enemy they forced the pace, but as soon as they approached they beat about, using days, weeks, and months before deciding to fight. Either side could refuse battle because they could get away while the other was arranging his lines." Frederick carried the operations of ancient war to the highest degree of perfection they ever attained.

Drill Not War

Among the innovations to be attributed to Frederick are: The division of armies so that they could march in a number of columns with less fatigue; the use of flank marches; the oblique order, a practical method of envelopment with the armies of his time; the lightening of the cavalry; increased mobility of artillery and the development of horse artillery; increase in the number of howitzers.

It has been one of the misfortunes of armies that Frederick's great reputation led to slavish imitation of the forms of the Prussian military system. Young officers from England and France attended the reviews at Potsdam and thought all the secrets of Frederick's success lay in Prussian drill, Prussian uniforms, and the spit-and-polish tradition. They were unable to distinguish the symbol from the substance in the Prussian army.

Drill was mistaken for the art of war although Frederick never so interpreted it. He "laughed in his sleeve," says Napoleon, "at the parades of Potsdam when he perceived young officers, French, English, and Austrian, so infatuated with the maneuver of the oblique order, which was fit for nothing except to gain a few adjutant-majors a reputation." (Quoted by Col. E. M. Lloyd, *A Review of the History of Infantry.)*

Napoleon ranked Frederick the Great with Caesar, Hannibal, Turenne, and Prince Eugene. What distinguished him above all other generals, Napoleon thought, was his extraordinary audacity. Frederick's fame was quickly eclipsed by that of Napoleon himself. The result has been that his battles have not been truly appreciated nor is there even in English any good analytical account of them. Guibert, in his *Essai General de Tactique,* was an admitted expounder of Frederick the Great's methods, just as Jomini later was to be of Napoleon's. Guibert influenced Napoleon profoundly. The introduction of a permanent divisional organization into the French army by the Duke of Broglie in 1759 made possible the new advances in mobility and rapid deployment of which Napoleon was to take such great advantage.

The Instruction of Frederick the Great for his Generals was written in 1747 following an illness. It is a remarkable book for a Prince, aged 35. It was revised in 1748 under the title of *General Principles of War.* One copy of it was sent to his successor to the throne in 1748 with a request enclosed that it should be shown to no one. In January, 1753, an edition of fifty copies was printed and sent to a list of officers whom the King considered models of their profession. A cabinet order enjoined each recipient on his oath not to take it with him in the field and exercise care that on his death it would be handed over to the King again, sealed.

General Czettertiz was captured by the Austrians in a small affair, February 21, 1760. On him was found Frederick's secret manuscript. It was duly prized, reprinted in German in 1761 and translated back into French and printed in France the same year. It was translated into English in 1762 under the title, *Military Instructions by the King of Prussia.* It was reprinted and edited by Scharnhorst [Prussian General, 1755-1813, who planned the reform of the army.] in Germany in 1794 for instruction purposes. It appeared again in the collected works of Frederick the Great in 1856 and in the military writings of Frederick the Great edited by General von Taysen in 1891.

The present text is translated from the *Instruction* of 1747, published for the first time in Berlin in 1936, on the one hundred and fiftieth anniversary of the death of Frederick the Great. This edition has been very carefully edited by Dr. Richard Fester and contains the original French text and a German translation. The present translation, made with permission of the German publisher, E. S. Mittler & Sohn, early in 1940, marks the first appearance in English of the Instruction of 1747.

Father Blazed Path

In his comments upon the father of Frederick the Great, Frederick William I, General William A. Mitchell* reminds us that it was the sire who laid the foundation for the power of Prussia under his son, by a wise economy, military severity, and the establishment of a formidable army. He reduced his court until it was the least expensive in Europe, and did away with luxury in every form; but he did not economize on his army. He spent four-fifths of his revenue upon it, and increased it from 38,000 to 84,000 men, of whom more than 26,000 were mercenaries. For a population of two and a half million this was a very large army. France with a population ten times as great and with eight times as much revenue had an army only twice as large.

Old Frederick William had a mania for enormous soldiers. He kept agents all over Europe, who recruited or kidnapped the tallest men that could be found, to swell the ranks of the regiment of giant guards at Potsdam. One monk was dragged from his convent and given 5000 florins for enlisting; a priest was pulled from his pulpit; an Irishman was abducted from Ireland and paid 1000 florins. These abductions caused many protests, but still they did not cease. It was not this regiment of giants, but the disciplined army and well filled treasury that did Frederick the most service when he came to the throne.

Frederick the Great knew military history and he was aware that Prussia's survival in the midst of the hostile countries of Central Europe depended solely upon its army. With this idea in mind, in the first year of his reign, he increased the Prussian army to 100,000 men, and incidentally, abolished the Potsdam Guard of Giants.

* In his *Outlines of Military History*, the Military Service Publishing Company, Harrisburg, Pa.

Strange to say, Frederick's army was composed largely of non-Prussians. The foreigners were a third and later a half of the entire army. Many methods were used to obtain recruits, including kidnapping. The pay of the soldier was only three half pence a day; and the food, clothing, and quarters were not good. The discipline caused many desertions. There was little patriotic sentiment in the army; it was a professional force which sometimes overleaped the bounds of discipline and gave way to pillage and merciless methods. Some soldiers were of such evil character that they could be controlled only by the strict discipline and severe punishments designed for that purpose.

Hatred of Troops

Personally, the soldiers hated Frederick and his officers, although they admired his courage and respected his abilities; but, in battle they would fight hard for him. The officers, drawn from a poor and numerous nobility, had little sympathy for the men, but their professional standard was high. They learned their business thoroughly, and promotion went by seniority and merit rather than by favor. The non-commissioned officers were excellent; most of them were promoted from the ranks and went no further; but some that came from the cadet schools rose to be officers. Thus, it may be seen that the Prussian successes were largely due to the ability of Frederick and his father in disciplining and training the officers and men.

Frederick reduced the firing line to three ranks. To increase the rate of fire, the skirts of the coat were cut away and the saber shortened so that men would not interfere with one another in ranks. Loading was accelerated by the use of iron ramrods, in place of wooden ramrods. Advancing troops could not load and fire as effectively as troops standing still; but Frederick trained the Prussians to keep up a steady and continuous advance, while firing volleys at intervals and reloading on the move, which was a sort of "marching fire." The men in the front rank fired with bayonets fixed; and Frederick later had all three ranks load and fire with fixed bayonets.

The effective fire superiority of the Prussians was followed up by a bayonet charge when close to the enemy. Frederick

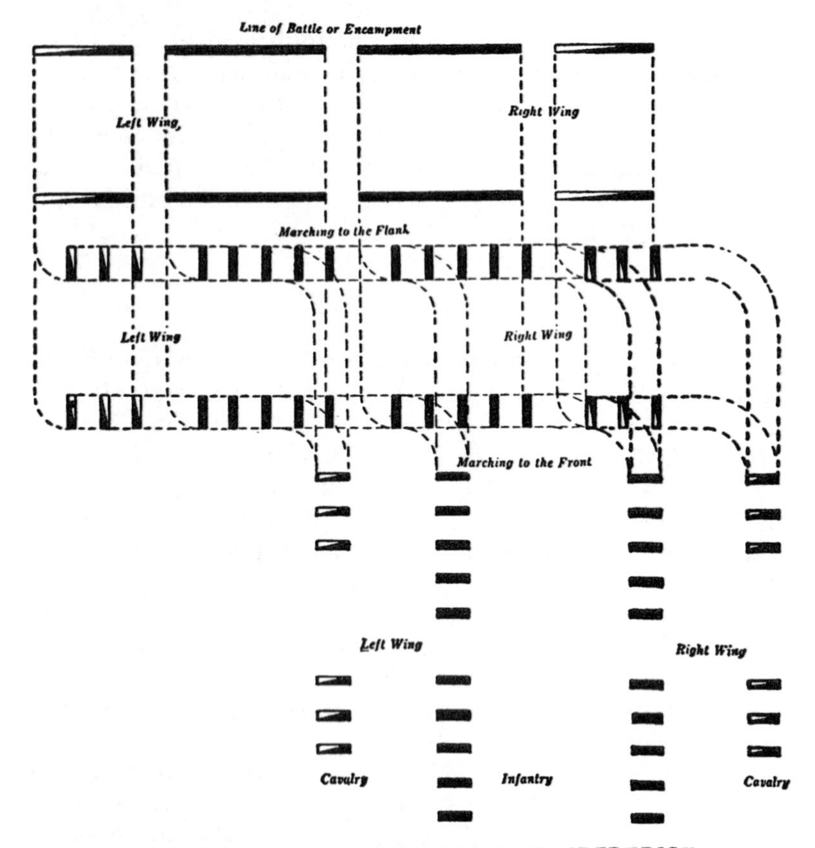

BATTLE FORMATION DEVISED BY FREDERICK

re-introduced marching in step, and gave great attention to preserving alinement and distance.

The infantry battalion consisted of six companies; but it formed for action in four divisions of two platoons each, regardless of companies. Column was formed from line by facing, for a short movement, and by turns of divisions or platoons for a long movement. In the presence of the enemy, movements were made in columns of divisions or platoons and line was formed: (1) by wheeling the divisions simultaneously to the flank; or (2) by halting the leading division and obliquing the others to the right and left and marching them to the front abreast the first.

Frederick preferred to wheel the divisions simultaneously to the flank when the flank of the column was not exposed to the enemy. The rapid and accurate formation of the battalion square was an important part of the drill. It was formed from line by wheeling the flank platoons.

The army was divided into four wings, two infantry wings in the center and two cavalry wings on the flanks. The army camped and fought in two lines, marched in two columns by the flank, and in four columns to the front.

Developed Oblique Order

As a result of these formations, Frederick developed a system of approach and attack which is called "Oblique Tactics." The idea was developed in the first two Silesian wars. In the battle of Hohenfriedberg, and still more in the battle of Sohr, (1745), the beginnings of the oblique order can be seen. The idea was not new, as it had been used by Epaminondas, a Theban general, in defeating the Spartans at the battle of Leuctra in 371 B.C.; Gustavus used the same principle at Lutzen. Frederick expressed his idea as follows: "You refuse one wing to the enemy and strengthen the one which is to attack. With the attacking wing you do your utmost against one wing of the enemy which you take in flank. An army of 100,000 men taken in flank may be beaten by 30,000 in a very short time."

The idea of the oblique order might be carried out by using two methods. The line of battle could be formed parallel to the enemy and the attack delivered in echelon beginning with greater concentration at one end of the line, headed by a picked body. This greater concentration on one flank was, in effect, the plan of Epaminondas; and, in open country, it was the only practicable method. But

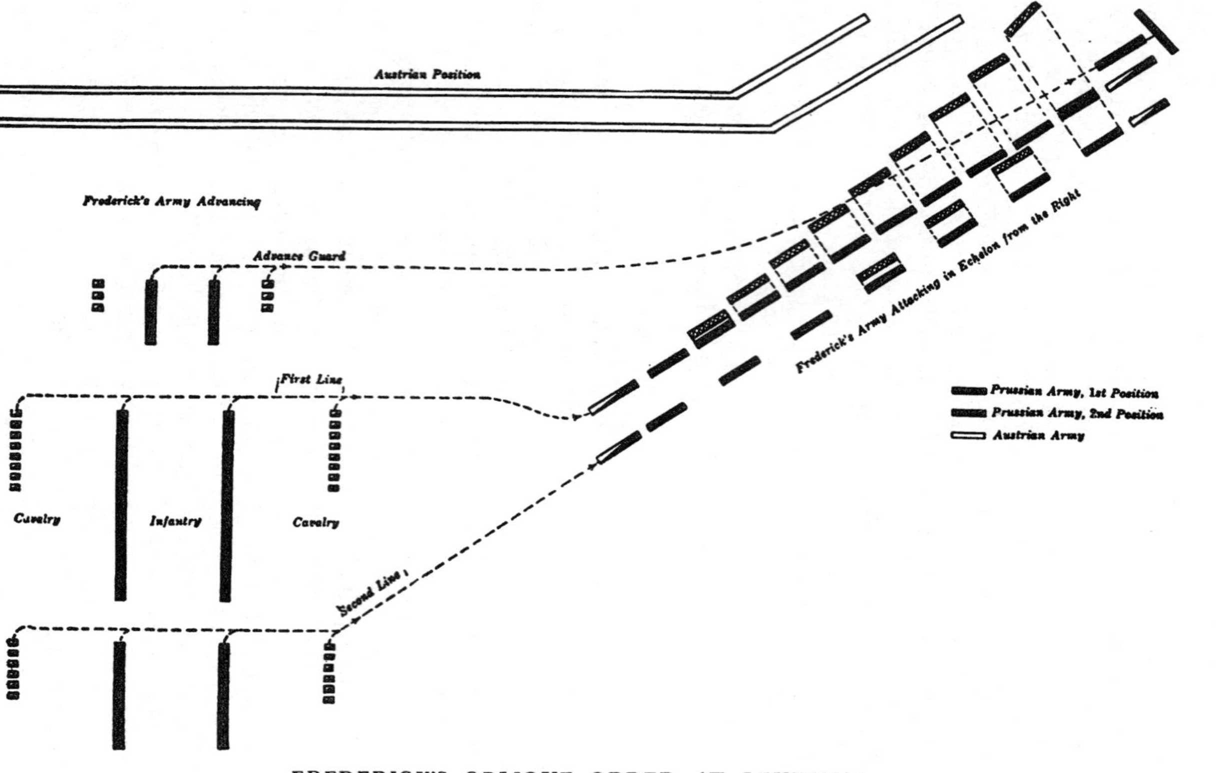

FREDERICK'S OBLIQUE ORDER AT LEUTHEN

where the ground, or darkness, afforded protection from observation and attack, the line could be formed at an angle to that of the enemy, overlapping his flank, which could then be rolled up.

At Leuthen, which is the one perfect example of the oblique order, Frederick combined the two methods. Under cover of some hills he placed his army on the Austrian flank and formed his line at an angle to the Austrian line. He then advanced in echelon from the right so that the front of the Austrians was struck blow after blow as the battalions came forward, at 50 pace distances.

The fame of the oblique attack rests chiefly on the victory at Leuthen. It was the first and only time that it was carried through against the enemy. As a war measure the oblique attack was dropped; but it was frequently exhibited later on the parade ground at Potsdam with the result that the officers and men believed that the oblique system had been the cause of their success.

"The oblique order of the Potsdam parades served no purpose but to make the reputation of certain adjutants," said Napoleon. The maneuver required great mobility, knowledge of the enemy's position, surprise, and inability of the enemy to reinforce the threatened flank or to attack the flank of the marching column. Frederick's oblique attack was practicable because his troops were excellent and his opponents very slow; but it was too stiff and inflexible and it depended too much on favorable conditions. Against Napoleon it broke down completely.

Improved Cavalry Tactics

Cavalry tactics were greatly improved in Frederick's armies. Mounted fire action was prohibited; and his cavalry was trained to charge in good order, boot to boot, increasing in speed as it neared the enemy, and to rely solely on the shock at full speed. The men charged with a yell to increase the moral effect. Pursuit was taken up after the enemy had been broken. In reforming, the cavalry rallied to the front

Musketeer *Fusileers* *Grenadier*

Dragoons *Hussar* *Cuirassiers*

SOLDIERS OF FREDERICK'S ARMY

instead of to the rear. Cavalry leaders were instructed to attack instead of waiting to be attacked.

Frederick's cavalry consisted of cuirassiers, dragoons, and hussars. His cavalry was distinctly Prussian, especially the hussars, who were counted on to bring back the men attempting to desert. The cavalry was at first formed in three ranks, and later in two. The dragoons could fight well on foot though they fought thus only when shock action was impossible. Frederick had two great cavalry commanders, Seidlitz and Ziethen, and the best fighting cavalry of the time. Frederick increased his cavalry from less than one-fifth to one-fourth the strength of his armies. It is said that out of 22 pitched battle fought by Frederick, his cavalry played the decisive part in fifteen.

Excellent as Frederick's cavalry was on the battlefield, it was inferior to that of the Austrians for the purposes of security and information. Croatia and other half-civilized countries on the Turkish frontier provided excellent light troops, among which the Pandours were conspicuous. The irregular Pandours and Croats formed a screen about the Austrian army; they reported every move Frederick made and kept him in ignorance of the Austrian movements. They could not defeat Frederick's cavalry in battle, but they always eluded it. They frequently captured Frederick's messengers and mail and on one occasion they completely cut his army off from the world and nearly starved it to death. Berlin was actually captured by a raiding column of 4,000 Austrian cavalry. They were excellent mounted troops for strategical purposes, as Frederick's cavalry was excellent for tactical purposes.

In Frederick's earlier campaigns artillery was almost neglected. Although guns and carriages had been lightened, artillery was still so heavy as to be of little use on account of its immobility. The horses were kept under cover and the guns were moved by hand on the battlefield. The heavy batteries of position were placed on the wings or in front of the line; and 2- to 4-pound battalion guns were placed between battalions or in front of them, under a corporal.

The corporal had no initiative and the battalion commander usually forgot to give orders about the guns. If the battalion was driven back, the guns were lost as a matter of course.

Frederick soon saw what excellent support his artillery could give his infantry, which was depleted during the wars; and he turned his attention to its improvement. At Leuthen, the Prussian battery that broke down the abatis on the Austrian left and raked its lines when they changed front, contributed greatly to the victory. At Hochkirch, Frederick's artillery saved his arm from utter destruction and sacrificed itself to cover the retreat. The artillery never had a chief of suitable rank and ability; and hence the Prussian artillery was never satisfactory.

Permanent Staff Formed

During the Seven Years War, Frederick and other generals developed a permanent staff, which was the forerunner of the modern general staff. Frederick was his own chief of staff and depended but little on his generalquartiermeister. He prepared his own plans, and needed only a body of adjutants to copy and distribute his orders. This body grew at the expense of the quartermaster-general's staff. Young officers, detailed from their regiments, were trained in engineering, topographical, and reconnaissance duties. As a result of this specialization, the staff acquired a degree of permanency. Frederick had field bakeries, as had the Austrians, and cooking wagons or kitchens. He organized magazines and supply depots at important centers, so that the troops did not have to plunder the country in order to live. He had a body of spies all over Europe, which rendered regular reports.

One of the most notable points about Frederick's infantrymen was their great fire power. This was due to their constant drill, and to the use of the iron ramrods. Dessau in 1698 provided his entire regiment with iron ramrods; and 20 years later they were supplied to all the Prussian infantry. The Prussians could fire six rounds a minute in drill, although it is to be doubted whether they could fire more than four in the field. The Austrians, who had wooden ramrods, could not fire more than two rounds a minute. Prussian soldiers carried thirty rounds. After Mollwitz, where they ran short of ammunition, an additional thirty rounds was carried in wagons and issued before battle. Each soldier had a piece of leather to protect his hand from the heat of the barrel. The range of the musket was under 200 yards. At close range the volleys of the Prussians were terribly effective; and the bayonet charge clinched the victory.

The Prussian cavalry was armed with the sword and carbine; only the cuirassiers wore breast armor. Frederick's artillery consisted of 3, 6, 12, and 24-pound guns, and 7, 10, and 25-pound howitzers and mortars. The range with solid shot was from 1,500 to 2,000 yards and with canister, 400 yards; the ranges varied with the type and caliber of the gun.

I

Frederick's Instructions, 1747

THE discipline and the organization of Prussian troops demand more care and more application from those who command them than is required from a general in any other service.

If our discipline aids the most audacious enterprises, the composition of our troops requires attentions and precautions that sometimes call for much trouble. Our regiments are composed half of citizens and half of mercenaries. The latter, not attached to the state by any bonds of interest, become deserters at the first occasion. And since the numbers of troops is of great importance in war, the general should never lose sight of the importance of preventing desertion. He can prevent it:

By being careful to avoid camping too close to large woods;

By having the soldiers visited frequently in their camps;

By keeping them busy;

By forming carefully a chain of guards around the camp so that no one can pass them;

By ordering patrols of hussars to watch the flanks and rear of the army;

By examining, when desertion occurs in a regiment or in a company, whether it is not the fault of the captain, if the soldiers have received the pay and comforts which the King provides for them, or if the officer is guilty of embezzlement;

By observing strictly the orders that soldiers shall be led in ranks by an officer when they go to bathe or forage;

By avoiding night marches unless they are required by the exigencies of war;

By careful observance of order on marches with strict prohibition against a soldier leaving the ranks or his squad, by placing officers at the debouches of defiles or where roads traverse the route of march, and by having hussars patrol the flanks;

By not withdrawing guards from villages until the whole army is under arms and ready to commence the march;

By hiding carefully from the soldiers the movements we are forced to make to the rear or by endowing retreats with some specious reason which flatters the greed of the soldier;

By preventing pillage, which is the source of the greatest disorders.

Capabilities and Particular Merit of Prussian Troops

The greatest force of the Prussian army resides in their wonderful regularity of formation, which long custom has made a habit; in exact obedience and in the bravery of the troops.

The discipline of these troops, now evolved into habit, has such effect that amidst the greatest confusion of an action and the most evident perils their disorder still is more orderly than the good order of their enemies. Consequently, small confusions are redressed and all evolutions made promptly. A general of other troops could be surprised in circumstances in which he would not be if commanding Prussians, since he will find resources in the speed with which they form and maneuver in the presence of the enemy.

Prussians' discipline renders these troops capable of executing the most difficult maneuvers, such as traversing a wood in battle without losing their files or distances, advancing in close order at double time, forming with promptness, reversing their direction suddenly to fall on the flank of the enemy, gaining an advantage by a forced march, and finally in surpassing the enemy in constancy and fortitude.

Obedience to the officers and subordination is so exact that no one ever questions an order, hours are observed exactly, and however little a general knows how to make himself obeyed, he is always sure to be. No one ever reasons about the possibility of an enterprise, and, finally, its accomplishment is never despaired of.

The Prussians are superior to their enemies in constancy since the officers, who have no other profession nor other fortune to hope from except that of arms, animate themselves with an ambition and a gallantry beyond all test, because the soldier has confidence in himself and because he makes it a point of honor never to give way. Many have been seen to fight even when wounded since the organization in general, proud of its past brave engagements, considers that any soldier who has shown the least cowardice in action disgraces it.

II

Projects of Campaign

WE cannot have any enemies except our neighbors, that is Austrians, whom I place at the head of them all; Saxons, the most envious of our expansion, and Russians, who can become our declared enemies as a consequence of the rivalry they have conceived since we have labored to relieve Sweden from their tyranny.

One should know one's enemies, their alliances, their resources and the nature of their country in order to plan a campaign. One should know what to expect of one's friends, what resources one has and foresee the future effects to determine what one has to fear or hope from political maneuvers. Since all these cases can be complicated and since it is impossible to foresee all the combinations that the caprice of fortune may bring, I shall embody my precepts in some general maxims.

One should not concern oneself in projecting a campaign with the number of the enemy, provided they do not exceed you more than a third. That is to say, with 75,000 men you can attack 100,000 in all security and even promise yourself to defeat them.

Extravagant projects of campaign are worthless. I call extravagant those which require you to make penetrations, or those which reduce you to a too rigid defensive. Penetrations may be worthless because in pushing too far into the enemy's country you weaken yourself, your communications become difficult to maintain on account of their length, and in order to make assured conquests it is necessary always to proceed within the rules: to advance, to establish yourself solidly, to advance and establish yourself again, and always to prepare to have within reach of your army your resources and your requirements.

Projects of absolute defense are not practicable because while seeking to place yourself in strong camps the enemy will envelop you, deprive you of your supplies from the rear and oblige you to lose ground, thus disheartening your troops. Hence, I prefer to this conduct the temerity of the offensive with the hazard of losing the battle since this will not be more fatal than retreat and timid defensive. In the one case you lose ground by withdrawing and soldiers by desertion and you have no hope; in the other you do not risk more and, if you are fortunate, you can hope for the most brilliant success.

Campaign Tactics

The projects of campaign that I propose for the offensive and the defensive according to my method are the following:

For the offensive I require a general to examine the enemy's frontier; after having weighed carefully the favorable and difficult factors in the different points of attack, he determines the locality through which he will advance. To make this more clear I am going to apply it to Saxony, Bohemia and Moravia.

To attack Saxony, I would assemble my army in the vicinity of Halle. The first thing to think about then will be subsistence; without supplies no army is brave, and a great general who is hungry is not a hero for long. The handling of food supplies should be entrusted to a man of integrity—capable, intelligent and discreet. It is essential to collect as much grain as it is thought will be consumed during a campaign, to protect the city where the food depot is established with a good garrison and secure it from any surprise, and even take precautions so that the enemy cannot have it set on fire by spies or hired agents.

This done, it is necessary to provide for the wagon trains. First you should take with you a large enough provision of grain to last for three weeks. Then, if the enemy is in the field, it is necessary to find out where he is and get rid of him. Following this, the seige of Wittenburg becomes indispensable; this makes you master of the Elbe and gives you communication with your country and covers it at the same time. From there it is necessary to march to the capital, which will fall of itself, expel the enemy from the country and collect supplies in winter magazines. The troops should not be dispersed too much. Your succeeding projects will depend on the circumstances of the time.

Campaign against Bohemia should be made only after

PROJECTS OF CAMPAIGNS

Here are illustrated Frederick's proposals for offensive and defensive campaigns against Saxony, Bohemia and Moravia and against the Austrians should the latter seek to invade Silesia and Brandenburg. (See text.)

much mediation. Examine the frontier! You will see four main passages, one beside the Lusace, another which leads to Trautenau, the third to Braunau, the fourth by the Comte of Glatz, Rückers, and Reinerz to Königgrätz. The one closest to the Lusace is of no value because you have no strong point in your rear, because you enter the kingdom in the corner of a difficult and mountainous country, where if the enemy happens to be there, he will have a wonderful chance to wage a war of ruses and chicanery against you.

The road to Trautenau is almost equally bad. If the enemy is found at Schatzlar, he will make your operation difficult, and perhaps impracticable, because of the advantage the terrain gives him and the heights which dominate all this country.

The road through Braunau is the best of all because you have Schweidnitz behind you. This is and will be made a good fortress because from there to Braunau it is only four miles and because, in spite of the difficulty of the roads, it is, nevertheless, the best of all those leading into Bohemia.

I prefer this to the routes of Glazt and Reinerz because in entering by Braunau you cover all of lower Silesia, the replenishment of the depots can be made with facility from Schweidnitz, while at Glatz this is very difficult because of the length of transport and the difficult roads. For in Silesia you should always regard the Oder as the nourishing mother for the magazines. It is closer than Schweidnitz and if there were only this reason, it should be regarded as decisive.

After having chosen this point of attack it is necessary to consider the security of the depots and the country. For this purpose we cannot dispense with sending a body of 6000 or 7000 men to the border of Neisse to oppose continually the incursion of the Hungarians and to cover the country in such a manner that the convoys from the interior of lower Silesia, which furnish and refill the depot at Schweidnitz, can arrive there in security. This upper Silesian corps has three points of support. One is Neisse, and for all operations beyond the Oder, Cosel and Brieg can serve for a retreat and depots.

Circumstances Control

As for the nature of your operations it is very difficult to determine them without knowing the circumstances in which you will find yourself. I daresay, always with certitude and from experience, that Bohemia will never be taken by making war there. In order to capture it permanently from the house of Austria it is necessary for an allied army to go along the Danube, while ours traverses Moravia, and for the two armies to arrive at the same time against Vienna, while a small body of troops cleans up Bohemia and draws contributions from it.

If you make war on the Queen of Hungary alone without having a marked superiority over her, your projects of campaign can only be a mass defensive, clothed with the externals and appearances of the offensive. The campaign will penetrate into Bohemia. You will take Königgrätz and even Pardubitz there, if you wish; but with these two cities you will have gained nothing. They are poor places in which to establish secure depots, because there is no tenable city nor navigable river in this country.

As a consequence there is no way to sustain yourself there during the winter without risking the troops, unless allies furnish you the means, as happened to us during the winter of 1741 and '42. Besides, the difficulties confronting convoys in wooded country, where they necessarily have to traverse the gorges of the mountains, always render your operations risky and exposed to the capture of your convoys by the enemy's light troops. Thus the campaign in Bohemia with equal forces against the Queen of Hungary will reduce itself to making part of your army subsist at her expense, after which your principal care should be to forage upon and exhaust all the outskirts of Bohemia which border on Silesia, so that the enemy will not be able to put large bodies in winter quarters there. These troops would molest you in Silesia, and your men would have no rest.

I would form an entirely different project of campaign against Moravia. It can only be attacked by the road from Troppau to Sternberg, or by the Prerau route. The Sternberg road is the closest, most convenient to Neisse, and, as a consequence, the more suitable. I am always assuming in these projects that the forces are approximately equal.

In this case it is necessary to leave a body of 7000 or 8000 men on the border of Braunau to oppose incursions that the hussars would be able to make from Bohemia into Silesia. If this corps found itself in contact with too formidable an enemy, there is a sure retreat at Schweidnitz. I am obliged still to make a second detachment in the vicinity of Jablunka, even more necessary than the first.

Firm Command Needed

The success of my whole project is founded on the firmness of the conduct of the officer who will command it. This officer should defend the entry into Silesia against the Hungarians, and this is why his defensive is important: If some clever and pillaging troops penetrate upper Silesia, the security of no convoy can any longer be counted on, and, just as the first campaign proposed in Moravia depends on subsisting on the Silesian depots, it follows that the convoys which are to follow the army should be able to do so with complete security.

As for magazines, I should establish my principal accumulation at Neisse, and advance a depot for two months' supply of grain at Troppeau. I should repair the walls, raise earthworks where needed, and pallisade the places which require it. From there I should march on Olmütz and fortify Sternberg in the same way to assure the passage of my convoys. Then I should attack the enemy wherever I found him; I should take thirty 12-pounder cannon, twenty-five howitzers and six 24-pounder pieces of light seige artillery, which will suffice to conduct the seige. The place will hold at least twelve days with open trenches. I should have my provisions advance there from Troppeau. I should have the breaches repaired; then I would advance on the enemy who will have withdrawn from the vicinity of Brünn.

It is then that it is necessary to employ ruses to draw him away from that place, on to the plains, in order to fight or finally to make ready to form the second seige. The fortress could hold eight days; if the enemy is in the vicinity, it will be necessary to build a strong entrenchment, to amass food there from Olmütz for three weeks, and to take the place. The city will not be able to resist a long time; it may be able to hold you for twelve days. This requires that the magazines follow, and as soon as you establish them at Brünn, as soon as the place is supplied, you can advance on Znaim and Nikolsbourg. This will throw the enemies back into Austria if they are thoroughly defeated and if they have not received help by this time. Even though they have left the field, the country is too favorable for them to expect that they will not send their light troops back on your two flanks, that is in the Trebitsch Mountains on your right and in the mountains near Hradisch on your left.

III

Defense To Offense

I AM now coming to what I call the defensive which turns into the offensive.

The greatest secret of war and the masterpiece of a skillful general is to starve his enemy. Hunger exhausts men more surely than courage, and you will succeed with less risk than by fighting. But since it is very rare that a war is ended by the capture of a depot and matters are only decided by great battles, it is necessary to employ all these means to attain this object. I shall content myself in exposing two projects of the defensive according to this method—one for lower Silesia, the other for the Electorate of Brandenburg.

I should defend lower Silesia against the Austrians if they intended to attack through Bohemia in the following fashion: I should establish my principal magazine at Schweidnitz, which I should garrison with five battalions and three squadrons of hussars. I should establish a small depot at the chateau of Liegnitz, to be prepared to follow the enemy if he went along the mountains. I should place seven battalions at Glatz with three or four regiments of hussars. The more of them that are there the better it will be. All the attention of the governor of Glatz should be directed on convoys and the depot of the enemy. If he strikes a good blow in that direction their projected offensive is ruined for a whole campaign.

On my part I should encamp the army at Schönberg, and throw up trenches so as to appear in fear of attack. If the enemy goes along the mountains, I should follow him, always taking up strong camps, but open, nevertheless, on the flanks and in the rear, if I were able. It would be necessary to fall either on his advance guard or on his rear guard and cut him off from it. Such a check is capable of reducing him to the defensive. If he acts as if he wished to attack me, he would find so many disadvantages in his situation that the desire will leave him.

If he wished to envelop me, it would be necessary to let him pass and to place myself behind him, upon which I would cut him off from his supplies. I would be able to choose ground of advantage to my troops and oblige him to fight in the locality that pleased me, and if I defeat him he will have no retreat and his entire army is lost. And I risk nothing by seeing him break into the country because my magazines are protected by my fortresses and he will be unable to capture them by a sudden blow.

Planning Rear Attack

I should make a project of defensive campaign for Brandenburg in the following manner.

I should assemble my army not far from Brandenburg. The enemy could only form two plans to penetrate the country. The best would be that of following the course of the Elbe; but since they would find Magdebourg, which is a fortress hard to digest, in their road, they would not think of this plan too readily.

Their plan would be without doubt, to assemble their forces around Wittenberg and to march straight on Berlin. In this case I should approach from the direction of Teltow, having my magazines at Brandenburg and Spandau. If the enemy had his supplies following in wagons, I should send all my light troops through the forest of Beelitz which extends into Saxony, and from there they would fall on the rear of the enemy and capture his convoys. The hussars would have a sure retreat in these great forests where they know the trails and through which they could always return to my army.

This defensive could only be made with vigor. It is essential to gain the rear of the enemy, or to surprise them in their camp, or to cut them from their country by a forced march. I would not advise anyone to place himself behind the Havel and the Spree; the country would soon be lost by this.

I shall limit myself to the two examples that I have cited. It is always necessary to form projects of campaign, as I have said, on estimates of the weather, and as this is always changing one cannot imitate in one season what has turned out well in another. I shall only add to these maxims that a general in all his projects should not think so much about what he wishes to do as about what his enemy will do; that he should never underestimate this enemy, but he should put himself in his place to appreciate difficulties and hinderances the enemy could interpose; that his plans will be deranged at the slightest event if he has not foreseen everything and if he has not devised means with which to surmount the obstacles.

[Jomini observes *(The Art of War)*: "During the first three campaigns of the Seven Years' War Frederick was the assailant; in the remaining four his conduct was a perfect model of the defensive-offensive. He was, however, wonderfully aided in this by his adversaries, who allowed him all the time he desired, and many opportunities of taking the offensive with success. Wellington's course was mainly the same in Portugal, Spain and Belgium, and it was the most suitable in his circumstances."]

IV

Subsistence and Commissary

UNDERSTAND that the foundation of an army is the belly. It is necessary to procure nourishment for the soldier wherever you assemble him and wherever you wish to conduct him. This is the primary duty of a general.

I divide the problem of subsistence into two parts, of which one deals with the place and manner of assembling supplies and the other with the means of rendering these magazines mobile and consign them to follow the army. The first rule is always to place magazines and fortified places behind the localities where you are assembling the army. In Silesia our principal magazine has always been at Breslau.

By placing your principal magazine at the head of your army, you risk being cut off from it by the first misfortune, and then you would be without resources; whereas, if your supplies were distributed by echelons the loss of one of these parts would not lead to that of the whole.

In the defense of the Electorate, one of your magazines should be at Spandau, the other at Magdeburg. The latter could serve for the offensive, and it is in relation to Saxony much like the depot of Schweidnitz to Bohemia and that of Neisse to Moravia. The first attention that should be given magazines is to choose the chief commissaries carefully. If they are dishonest men, the sovereign can lose prodigiously from theft; under any circumstances they should be watched carefully.

There are two methods of assembling magazines; one is to collect the grain in the country and credit the contributions of the peasants and gentlemen against their regular tax. It is not necessary to have recourse to the other expedient—that of contractors—except when the first method is physically impossible. These men plunder pitilessly and make their own law by the exorbitant prices they place on food.

I add again to these maxims that it is always necessary to form large magazines in good time, because a considerable period is required to amass supplies and sometimes the weather makes the roads bad, or rivers useless, or wagons impracticable. Besides the wagons attached to the regiments, the commissary should have enough vehicles to carry meal to subsist your army for a three weeks' period. These should be drawn by horses. Experience has demonstrated to us that oxen are of no use.

Transportation Stressed

Inspectors are required for the horses and the vehicles. The care that these men should have for their preservation is of the more consequence, for if they break down without your being able to replace them, your whole project will be halted. It is necessary for the general to keep his eye on all this.

Whenever an army operates along a river, subsistence becomes easier. The great generals, when they were able, have always reflected on this fact as being of great advantage. The Elbe gives us this advantage against the Saxons; but, if you wish to act in Bohemia or in Moravia, you can only count on your land transport. Sometimes three or four magazines are formed in file, from which you subsist, but as soon as you wish to progress into the enemy country, the magazine at Schweidnitz, for example, is fortified, while that at Breslau is obliged to refill in proportion as supplies are drawn from the first. The supply trains which follow the army are filled from this one. Then you are assured of supplies for four weeks.

In addition to the wagons, the field commissary carries with it iron ovens. A large number are necessary in order to have plenty of bread, for it is impossible to do without this. If you undertake an expedition, bread should be on hand for ten days ahead and biscuits for five. Biscuits are excellent in case of need, but in place of eating them like bread, our soldiers break them up and make soup of them, which is not sufficient for their nourishment.

I shall be asked: But if you advance, for example, into Bohemia with all this meal, where will you store it? I answer that this only can be done at Pardubitz, which is the only locality at all reasonably available. Otherwise storage must be in a city from which your army is supported and where you have your ovens. And upon advancing from there this mobile magazine always follows the army. I have had hand mills made for each company. These can be useful because wheat can be found everywhere and the soldiers can grind it. They then deliver the meal to the commissary who returns it to them converted into bread. This is a resource in necessity that experience should show whether more or less use should be made of it. But, I may be told, what will you do when you have lived four weeks in the enemy's country? Then your meal will be eaten. Before reaching that point I should return my caissons to the nearest depot and have them filled and conducted to the army.

This is the place to speak of convoys. The number of troops required to escort convoys increases or diminishes according to circumstances,—the remoteness or the nearness of the enemy, the strength of his light troops and the localities where he posts them. I have always made infantry the principal element of my convoys. This infantry is given some cannon and one hundred, two hundred or more hussars, simply to scout and notify the escort of the places where the enemy is.

The cities where the escort rests should be garrisoned with troops and in the same way, if there is some village located on difficult terrain, such as a mountain gorge, it should be occupied to facilitate the passage. The general also takes the precaution of putting out a detachment forward which covers the march of the convoy toward the most exposed flank. I cover the duties of the minor officers in my *Institutions Militaires.*

Beer, Brandy and Sutlers

When an invasion of enemy territory is planned, all the beer and brandy possible to furnish the army should be brewed on the frontier. And within the enemy territory all the breweries that are found near the encampment are put to work. It is necessary to protect the sutlers, especially in enemy country and send them out to pillage with the foragers in all the territory where the peasants have deserted their harvests. Under such circumstances the inhabitants do not furnish anything to the camp, and the sutlers have to pillage and supply it. In the camp itself, prices of everything are established with equity so that the soldier is not defrauded and the sutler is able to live.

I shall add to this subject: We have always furnished bread free to the soldier because, in making war in territory that the enemy had made into deserts, it was only just to relieve him. In addition he has been furnished two pounds of meat a week, from herds of cattle with us which followed the troops. This provision was supplied at the same time that the provision wagons came to camp.

Green and Dry Forage

I treat the question of forages here, since it has an affinity with magazines and the commissariat. Dry forage is collected in magazines. It consists of oats, barley, hay and straw which is cut. The oats should be dry and not sunburned, or they will disagree with the horses. Cut straw, according to many cavalry officers, bloats the horses without giving them any nourishment. These magazines serve to hinder the enemy in the field and forestall winter expeditions; however, it is difficult to move very far with the enormous wagon train that the nourishment of men and animals at the same time demands.

During the first campaign in Silesia the horses were fed with dry forage, but we had the Oder which carried the provisions up to Ohlau, only a short distance from where we were. When it is desired to make a winter expedition, hay for five days is sent out and each cavalryman is given oats for the same period to carry on his horse. Any time that it is desired to attack Bohemia or Moravia, it is necessary to wait until green forage is growing in the country, or the risk is run of being obliged to make the campaign without cavalry. Green forage can be seized in the field and the dry can be taken later in the season in the villages where the peasants have harvested their crops.

This is done by large foraging parties escorted by cavalry and infantry with numbers proportioned to those the enemy is able to oppose to us. Foragers are ordered out either by wings or by the entire army. They assemble by wing toward the flank where they wish to forage, and from the whole army, either in its front or rear. A foraging party is composed of the escort of cavalry foragers, those of the infantry and some of the army wagons.

If foraging is being done in level country, the hussars march at the head and move uncovered. They are followed by some squadrons and battalions mingled with the foragers, with a battalion between them of foragers from the infantry, and some cavalrymen and grenadiers who form the enclosure, followed by a rear guard of hussars. The foragers should be armed with their sword and a carbine.

When you arrive on the field which you have had reconnoitered and which you wish to forage, you first form your chain, that is to say you garrison the villages, if there are any, with infantry, placing the cavalry squadrons between them; if there are no villages, the cavalry should be mingled with the infantry, which always carries its cannon with it.

You retain a reserve of cavalry which you hold in the center, so as to be able to use it on any flank where the enemy wishes to make his greatest effort to penetrate. This disposition having been made, the ground is divided up and foraging is commenced. On such an occasion the hussars need only to skirmish to keep away the enemy. Care should be taken that the foraging is well done and the bundles are large. When everything is completed, the foragers return to camp with some escort. When almost everyone has left, you reassemble the escort and, forming the rear guard, follow the rest.

Foraging parties in villages are made about the same way with the difference that the infantry of the escort garrisons the enclosures of the village, or defiles in the road and the cavalry is placed on the wings, on the flanks, or sometimes in the rear of the village in accordance with the demands of the terrain. The most difficult of all foraging is in wooded and mountainous country. Then the advance guard and rear guard come from the infantry. Foraging parties are formed according to the occasion.

If you wish to remain in a camp and the enemy is in your vicinity, your foraging between his army and yours amounts to living at his expense. Following this you forage a league around, taking the most distant forage first, always approaching your camp and conserving the nearest to the last. If you only encamp for a few days you will forage what is easiest to bring in.

V

Importance of Camps

THE art of choosing his camp well is, according to my opinion, one of the primary studies that a general should make because of its importance and the influence of the camp on many details which are related to it. Those who are interested I refer to my *Institutions Militaires.* In this work I am concerning myself only with the larger aspects of war and the functions of the general. Camps are of different natures, and they are not perfect except to the extent that they meet the uses for which they are intended. There are camps to cover a country, offensive camps, defensive camps, entrenched camps for armies making sieges, foraging camps, rest camps and concentration camps. The camps which cover a country should be well chosen, primarily with knowledge of the country, which is the basis of projects of war, just as the checkered-board is indispensable to those who want to play chess.

I am first going to tell about three camps which equally cover a country. There is one near Neustadt in upper Silesia. You leave the city and river of Hotzenplotz in front of you and, by raising two redoubts on the wings of your army, you cover all of lower Silesia. This is the place where the camp should be, but, after having marked the place, the skill of the general decides the manner in which he occupies its ground.

The first rule is always to occupy the heights; the second, that if you have a river or a stream in front of the camp, not to move more than half a rifle shot's distance from it. Tallard and the Elector of Bavaria were defeated at Höchstadt for having neglected this rule.

The third: to place your camp so that if the enemy passes on your right or on your left, the terrain will give you an equal advantage. Bends in rivers, marshes, or precipices will serve for this.

The fourth: having your front and flanks well supported,

your rear should be free; that is to say that there are no defiles so you are able to make a march to the rear in several columns without embarrassment. In these unassailable sorts of camps you are able to garrison the villages with infantry. It is necessary that you have laid out in advance two other camps on your flank so that everything will be prepared and that you can act first in every eventuality. There is a camp between the cities of Neisse and Ottmachau just like this. Those who have a military eye for the terrain and knowledge of war will see it without me designating it more precisely.

On the borders of Bohemia there is a similar camp at Liebau. Two roads meet there. The enemy cannot turn you except by way of Braunau. The city of Liebau should be on your left. If you are there simply to cover the country, it is necessary to raise redoubts at varying distances in front of the infantry and to accommodate yourself to the ground. The fortified city of Schweidnitz and the Schweidnitzer Wasser which flows there will also furnish you with camps for the defensive and from which your skill can gain a great advantage.

The Electorate is more difficult, not to say impossible, to cover. It is a large front which is too long to be defended; in placing yourself near Berlin you abandon all the rest. In a word this country cannot be defended except by battles.

Entrenched Camps

Camps for the offensives are open in front, but the wings should always be covered by marshes, streams, woods, or precipices. I do not speak of villages at all, since in our locality they are almost all of wood and there is no way to defend them. However, in all camps, villages which are within a quarter of a mile from the front and flanks of the camp are garrisoned so that enemy detachments cannot sneak into them and approach the army too closely, and to secure it against surprise. However, they are almost never garrisoned during fights, as I shall explain in its proper place. It should be observed that your camp should always be withdrawn about three hundred paces from the true field of battle that you choose, so that in case of surprise and there will not be time to strike tents, your two lines will have room to form and maneuver.

There are rules to observe for entrenched camps surrounding besieged villages. First, the entrenchment should not be made too large or of such a nature that it cannot be garrisoned in a continuous line by your infantry. The terrain should be used carefully. Marshes, streams, or abatis which can narrow the extent of ground to be held should be put to profit. The least heights should be occupied to avoid having your entrenchments dominated.

Ditches should be as deep as possible. The entrenchments should be flanked with redans and redoubts. The weakest localities should be fortified with foxholes, or by some works, and even by fougasses. Food and provisions for the entire length of the siege should be massed in the camp. When the city is taken the entrenchments should be filled up.

Camps for defense are similar in many respects to those of the first type of which I have spoken. I should add that usually they are chosen either on high ground, such as that at Marschowitz, or behind defiles, or in the bend of some river, like that of the Prince of Lorraine, in 1745, near Königgrätz, or near fortified places, like that of Monsieur de Neipperg after the battle of Mollwitz. But it is not sufficient just to have found one camp; it is necessary that you should have two or three other sites in your head that you can occupy in your rear in case the enemy turns you.

Of all the countries I know, Bohemia is the most favorable for strong camps. You even are often obliged to use them when you are looking for others. I should add to this something that I cannot repeat often enough, which is to warn that when you place yourself in a strong camp you do not enclose yourself in a *cul de sac.* Thus I call a post unassailable, when you cannot leave it, except through defiles; for, if the enemy is clever and vigilant, he will seize the defiles, and in this case you will be taken without fighting. Your camp should always be open at the rear.

Foraging camps are sometimes close and sometimes distant from the enemy. I shall speak only of the first, the others demanding less application. Either foraging camps should be extremely strong, naturally, or they must be fortified, lest the enemy be able to attack you suddenly, while your foragers are absent, and profit from your weakness to cut you off. If you have no other purpose than to forage on enemy country, it must carefully be hidden from him; this is done by false demonstrations of which I shall speak in their place.

Guarding Against Surprise

This article is only concerned with the precautions to be taken so as not to be surprised while your forces are divided up to forage. To accomplish this it is necessary to choose infantry posts, place infantry in the first line, raise redoubts where the ground requires it, and make abatis where there are woods, much as we did in the camp at Chlum and the Prince of Lorraine in his strong camp at Königgrätz.

After these first precautions have been taken it is necessary to conceal the days and places where you are going to forage. One or two days after arrival in camp, the environs to a distance of a mile and a half should be reconnoitered under the pretext of waging guerrilla war. Guiding yourself on the information you receive, have all the different foraging parties ready, but do not issue the orders until late in the evening to the officer who is to execute them; send out a large number of small parties to obtain information of the movements of the enemy, and regulate yourself by that. Foraging is done in greater security when it is carried on during the same day and at the same time as that of the enemy.

Rest camps are those where there is nothing to be done and where the *demarches* of the enemy are awaited, or where one waits for spring to commence operations. The most tranquil are those which have rivers or marshes in front of them, and the most convenient are those which are closest to the magazines. Such was our camp at Strehlen.

The duty of the general in these sorts of camps is to reorganize his army and personally to oversee order and discipline. The cavalry and infantry should be visited often and examined to determine if the mounts are well nourished and if these two corps are complete. The energy of the general can often find means to recruit troops. A man with his heart in his profession imagines and finds resources where the worthless and lazy despair, for—I repeat it again—numbers are an essential point in war, and a general who loves his honor and his reputation will always take extreme care to conserve and recruit his troop.

Likewise, the troops should be exercised frequently, cavalry as well as infantry, and the general should often be present to praise some, to criticize others, and to see with his own eyes that the orders which are found in my *Institutions Militaires* are observed exactly. The camp guards should be visited to see if they are vigilant, if the officer is attentive or negligent, and if he is extremely strict so that the mounted sentries will not allow anyone to enter the camp without examining and announcing him, and to prevent desertion by all the means that I have indicated previously.

The army, divided into separate brigades, moves into a concentration camp that has been laid out under the orders of four or more lieutenant generals. Often they are allowed to camp in separate bodies to facilitate subsistence, but close enough, however, to be able to unite within a few hours.

But this sort of camp is never chosen close to the enemy, where one runs risks like the King of England at Dettingen. He camped by corps and marched in the same fashion when suddenly he found the French army on his road without expecting it. In all your camps, the water should be close at hand and good and there should be enough wood nearby so that the soldiers can cook their rations. In addition, to be secure against surprise by numbers of light troops in which the enemy is ordinarily superior to us, we have adopted the usage of the Romans of making light entrenchments all around our camp, as can be found in my *Institutions Militaires.* This also prevents desertion.

Infantry pickets surround the camp, front and rear, as is prescribed in my regulations. A light entrenchment is made surrounding the camps in which one expects to remain. The villages on the wings or a quarter of a mile in front are garrisoned with battalions of grenadiers, and sometimes battalions are placed in the posts which are beyond the extremity of the wings, where there are either bridges or causeways to pass, to cover the camp and to guarantee it against the incursions of light troops.

Measures to Protect the Camp

The cavalry camp guards are placed toward the front and according to the requirements of the situation; since they are only posted to warn of the enemy, large bodies are not used, at least unless you are close to the enemy army and nothing separates you. With an army of 40,000 men we have had but 300 troopers on guard and sometimes fewer, but other precautions are taken which are more effective than these grand guards to prevent surprise. I have given the regulations for placing them in my *Institutions Militaires.*

Detachments are pushed out in front or on the flanks of the army toward the enemy. This sort of detachment should be composed of cavalry, infantry and plenty of hussars. By this mixture of arms they sustain each other on all types of terrain. Such a corps, composed of two or three thousand hussars, one thousand cavalry, and fifteen hundred or two thousand grenadiers, is advanced a half or three-quarters of a mile ahead of the army. One of the most skillful and vigilant generals is chosen for this command. He should camp in a locality where he will have a defile or a small wood, of which he is the master, in front of him. In this case he lines the other side of the wood toward the enemy with his mounted sentinels and camps his corps on this side of it.

He should have patrols continually on the road, and messengers should be detached from them day and night to keep him informed of what is happening and to send notice to the general. If you see that numbers of the enemy's light troops attempt to cut the communication with your advance guard, you may conclude from this that the enemy is on the point of executing some enterprise, directed at either this corps or the army. Be on guard in this case and send a large detachment strong enough to force its way and give you notice of what is happening.

Even though a detachment or advance guard covers the camp, nevertheless have your hussars patrol on your wings and rear so that no precaution shall be neglected. I should add to these regulations that no general can be permitted to lodge in a village that is not situated within the camp itself.

VI

Study of Enemy Country

KNOWLEDGE of the country is to a general what a rifle is to an infantryman and what the rules of arithmetic are to a geometrician. If he does not know the country he will do nothing but make gross mistakes. Without this knowledge his projects, be they otherwise admirable, become ridiculous and often impracticable. Therefore study the country where you are going to act! When it is desired to apply oneself to this essential part of war, the most detailed and exact maps of the country that can be found are taken and examined and re-examined frequently. If it is not in time of war, the places are visited, camps are chosen, roads are examined, the mayors of the villages, the butchers, and the farmers are talked to.

One becomes familar with the footpaths, the depth of the woods, their nature, the depth of the rivers, the marshes that can be crossed and those which cannot, and one thus learns to distinguish carefully between the conditions of marshes and streams in different seasons of the year. It cannot be said that a stream is impracticable in the month of August because it had been in the month of April. The road is chosen for such or such a march, the number of columns in which the march can be made estimated and all strong camping places on the route are examined to see if they can be used.

In reconnoitering these camps they are examined to determine if exits are numerous and easy, if they can be fortified by damming streams and about what effect can be expected from this; if abatis can be made, and how large a number are required for this work to complete it in a certain time. Following this, the plans open to the enemy are considered, the marches he might make and the camps he might occupy are examined, and an estimate is made as to whether he could be attacked during his march or in his camp, he could be turned, and all movements that could be made against him are studied. The larger cities and the better [walled] cemeteries on the outskirts are reconnoitered to estimate likewise to what extent they could serve the one who occupies them.

This sort of knowledge can be obtained promptly in the plains, but requires more skill and research in mountainous country where one of your first cares is to place your camp so that it will not be dominated, but be dominant over all the surrounding ground. The difficulty of procuring this advantage will perplex you, for you may be hindered in obtaining water, which must necessarily be close to camp, and you dare not place your troops in a locality more distant from your road than they should be.

Assume that you wish to form a project of campaign against Moravia. Examine carefully the best map that you can find and you see the principal highways which lead into this country, one from Glatz through Mittelwalde, Landskrom, Littau, and Olmütz; a second through Hof, Bentz, Freudenthal, Sternberg, and Olmütz; the third through Hultschin, Ostrau, Fulnek, Weisskirchen, Prerau, and Olmütz. These different routes should be ridden over, and defiles and posts from which the enemy could profit, the means of evading his opposition, the number of columns with which you will be able to march, and whether certain passages are not excessively bad, should be noted especially.

It is necessary to learn if there is no other way by which you can avoid an obstacle that appears invincible. Make a detour and judge if, by taking it with the army, the enemy would be able to prevail against this movement, or if you would be able by this means to place him in a difficult position. When you reconnoiter these three routes and have examined the locale with all your attention, it is necessary to see Olmütz.

I assume that you have a map of it and that you will always estimate your projects, and what the enemy may be able to do, in this fashion. But in such reasoning it is essential to be objective and it is dangerous to delude oneself. To accomplish this put yourself in the place of your enemy, and all the hindrances which you will have imagined and which he will not make for you, when war comes, will be just so many things that will facilitate your operations. Good dancers often go through their steps in sabots and they become more agile when they are in pumps. An examination of this nature needs to be made with reflection. As much time should be allotted to it as required; when this is done in a neighboring country, it is necessary to hide your secret intention with the most specious pretexts that you are able to invent.

Value of Quick Comprehension

The ability of a commander to comprehend a situation and act promptly is the talent which great men have of conceiving in a moment all the advantages of the terrain and the use that they can make of it with their army. When you are accustomed to the size of your army you soon form your *coup d'œil* with reference to it, and habit teaches you the ground that you can occupy with a certain number of troops.

Use of this talent is of great importance on two occasions. First, when you encounter the enemy on your march and are obliged instantly to choose ground on which to fight. As I have remarked, within a single square mile a hundred different orders of battle can be formed. The clever general perceives the advantages of the terrain instantly; he gains advantage from the slightest hillock, from a tiny marsh; he advances or withdraws a wing to gain superiority; he strengthens either his right or his left, moves ahead or to the rear, and profits from the merest bagatelles.

This also is required of the general when the enemy is found in position and must be attacked. Whoever has the best *coup d'œil* will perceive at first glance the weak spot of the enemy and attack him there. I shall have occasion to extend my observations concerning this in the article on battles. The judgment that is exercised of the capacity of the enemy at the commencement of a battle is called *coup d'œil.* This latter is the result only of experience.

Exact knowledge of the terrain regulates the dispositions of the troops and the order of battle. Our modern formations for combat, for the most part, are defective because they all are cast in the same mold: the infantry in the center and the cavalry on the wings. If an army is really to be disposed according to the rules of war it is essential that each arm should be placed in the locality where it can act. Infantry is most suitable for outposts; cavalry should always be in the plains or it becomes useless and cannot act. Never encamp your cavalry near woods of which you are not the master, nor near impassable marshes, nor in ravines parallel to your camp which prevent your troops from acting. For if the enemy intends to attack you and perceives your mistake, he will utilize these advantages.

Cavalry and Terrain

He will oppose your cavalry with infantry and cannon, and he will rake them with rifle fire to the point that, dismayed at being killed uselessly, they take to flight. Therefore, without worrying about ordinary tricks, put all your cavalry on one wing! Put it all in the second line! Divide it equally between the two wings, or place it without observing the order of battle, but according to the terrain that permits it to act.

As for Prussian infantry, it is superior to all rules. However, open country suits it best; I repeat it again. Never pen it in villages and do not choose camps which are merely unassailable, but open enough so that you can attack. The power of the Prussians is in the attack. Farther on, when I speak of battles, I shall give more illumination on this point. I observe that, regardless of circumstances, a corps of your army should always be destined for the reserve, even when you camp on two lines. The regiments which you intend to use for this purpose should be notified in advance. I shall have occasion to pluck this string in the article on battles, for rearguards are the safety of armies and often they carry victory with them.

VII

Detachments; How and Why Made

THERE is an ancient rule of war that cannot be repeated often enough: hold your forces together, make no detachments, and, when you are ready to fight the enemy, assemble all your forces and seize every advantage to make sure of success. This rule is so certain that most of the generals who have neglected it have been punished promptly.

Thus the great Prince Eugene was defeated at Denain because he did not have time to come to the support of Lord Albemarle's detachment. Guido Stahremberg also was defeated at Almanza through the fault of the English, who had failed to join him. The Prince of Hildburghausen endured a similar affront at Banjaluka and Wallis on the banks of the Timoc, for being detached from the Imperial army, and finally the Saxons at Kesseldorf for not having called upon Prince Charles of Lorraine, who was only a day's march distant. Likewise, we could have been defeated at Soor, if the exceptional courage of the troops and the skill of the generals had not pulled us out of the affair.

Hence the subject of detachments is extremely delicate. None should be made, except for good reasons, if you are acting offensively in open enemy country and are only master of some strong point. If you are actually waging war never throw out detachments except for convoys. In countries like Bohemia and Moravia, which are very mountainous, you will be obliged to leave detachments to guard the mountain gorges through which your convoys arrive, until the time when you have established your magazines in a fortified city. In such a case there are two precautions to take: first to choose strong encampments; the other, to place even these detachments where they are able to defend themselves in any situation until you can come to their help.

I do not call an army corps which is used as an advance guard a detachment, and as for other detachments, it is necessary to provide a secure retreat for them. I did this for ours in Upper Silesia, where, in case the enemy was too strong for them, they had a secure retreat in the fortresses of Neisse, Brieg, or Cosel. Officers who command detachments should be determined men, intrepid and prudent. Light troops should never be able to disturb them, but at the unexpected approach of a large army corps they must look out for themselves, and they should know how to withdraw before a superior force and to profit in turn by advantage of numbers.

Ordinarily, most detachments are made in defensive wars. Petty geniuses attempt to hold everything; wise men hold fast to the most important points. They parry great blows and scorn little accidents. There is an ancient apothegm: he who would preserve everything, preserves nothing. Therefore, always sacrifice the bagatelle and pursue the essential! The essential is to be found where big bodies of the enemy are. Stick to defeating them decisively, and the detachments will flee by themselves or you can hunt them without difficulty. Thus we abandoned Upper Silesia to the pillage of the Hungarians, and we hastened with all our forces to oppose the Prince of Lorraine; once he was beaten, Nassau without difficulty purged Upper Silesia of the Hungarians who infested it.

VIII

Talents of a General

A PERFECT general, like Plato's republic, is a figment. Either would be admirable, but it is not characteristic of human nature to produce beings exempt from human weaknesses and defects. The finest medallions have a reverse side. But in spite of this awareness of our imperfections it is not less necessary to consider all the different talents that are needed by an accomplished general. These are the models that one attempts to imitate and which one would not try to emulate if they were not presented to us.

It is essential that a general should dissemble while appearing to be occupied, working with the mind and working with the body, ceaselessly suspicious while affecting tranquility, saving of his soldiers and not squandering them except for the most important interests, informed of everything, always on the lookout to deceive the enemy and careful not to be deceived himself. In a word he should be more than an industrious, active, and indefatigable man, but one who does not forget one thing to execute another, and above all who does not despise those little details which pertain to great projects.

The above is too vague; I shall explain.

The dissimulation of a general consists of the important art of hiding his thoughts. He should be constantly on the stage and should appear most tranquil when he is most preoccupied, for the whole army speculates on his looks, on his gestures and on his mood. If he is seen to be more thoughtful than customary, the officers will believe he is incubating some project of consequence. If his manner is uneasy, they believe that affairs are going badly, and they often imagine worse than the truth. These suppositions become army rumor and this gossip is certain to pass to the enemy's camp.

It is necessary, therefore, that the personal conduct of the general should be so well reasoned that his dissimulation will be so profound that no one can penetrate it. If he fears that he cannot master his bearing, either he can pretend to be ill or he should make an excuse of some personal trouble to explain his disquieted appearance. Above all, when he has received bad news, he should treat it as a trifle and show the number of resources which he has to counteract it.

Respect For the Enemy

While never despising his enemy in his heart, he should never speak of him except with scorn and should compare carefully the advantages of our troops over the others. If some detachment is unfortunate, he should examine the cause; and, after having determined the reason for the fault, he should instruct his officers concerning it. In this fashion a few minor misfortunes will never discourage the troops, and they will always preserve confidence in their ability.

Secrecy is so necessary to a general that the ancients have even said that there was not a human being able to hold his tongue. But here is the reason for that: If you form the finest plans in the world but divulge them, your enemy will learn about them, and then it will be easy for him to parry them. The general plan of the campaign should be communicated at most to the officer responsible for supplies, and the rest of the details should not be told to officers except when the time has come to execute them. When there are generals detached and you must write to them, the letter should be completely in code. If the enemy intercepts the message you will not have betrayed yourself.

Since there are prodigious preparations to be made for war and some must be started early, secrecy may be betrayed thereby. But in this case you must deceive your own officers and pretend to have designs which you want the enemy to attribute to you. He will be notified by the indiscretion of your officers, and your real intention will remain hidden. For surprises and sudden blows, well-thought-out instructions should be prepared, but they should not be delivered to officers until the moment of execution.

When you are planning to march, arrange everything in advance, so as to be able to act freely, always under other pretexts, and then do suddenly what you have proposed to yourself. It is absolutely necessary to change your methods often and to imagine new strategems. If you always act in the same manner your methods soon will become known, for you are surrounded with fifty thousand curious who want to know everything that you think and how you are going to lead them.

Attitude Toward Troops

The commander should practice kindness and severity, should appear friendly to the soldiers, speak to them on the march, visit them while they are cooking, ask them if they are well cared for, and alleviate their needs if they have any. Officers without experience in war should be treated kindly. Their good actions should be praised. Small requests should be granted and they should not be treated in an overbearing manner, but severity is maintained about everything regarding the service. The negligent officer is punished; the man who answers back is made to feel your severity by being reprimanded with the authoritative air that superiority gives; pillaging or argumentative soldiers, or those whose obedience is not immediate should be punished.

The general even can discuss the war with some of his corps commanders who are most intelligent, and permit them to express their sentiments freely in conversation. If you find some good among what they say, you should not remark about it then, but make use of it. When this has been done, you should speak about it in the presence of many others, it was so-and-so who had this idea; praise him for it. This modesty will gain the general the friendship of thinking men, and he will more easily find persons who will speak their sentiments sincerely to him.

The principal task of the general is mental, involving large projects and major arrangements. But since the best dispositions become useless if they are not executed, it is essential that the general should be industrious in seeing whether his orders are executed or not. He should select his encampment himself, so that thereby he may be the master of his position, that the plan of it may be profoundly impressed on his mind, that he may place the cavalry and infantry guards himself, and that he may order on the spot the manner in which villages should be occupied. Afterwards, no matter what happens, he is able to give his orders with knowledge, and to make wise dispositions from his information of the ground. The more that all these minor details are well thought out, the more he will be able to estimate what the enemy is able to do and the more tranquil he will be, having foreseen everything and provided in his mind for everything that might happen to him. But all this is not done except by his own energy.

Thus be vigilant and indefatigable, and having made one tour of your camp, do not believe that you have seen everything. Something new is uncovered every day, and sometimes it is only after having reflected two or three times on a subject that useful ideas come to us.

Value of Skepticism

Skepticism is the mother of security. Even though fools trust their enemies, prudent persons never do. The general is the principal sentinel of his army. He should always be careful of its preservation and see that it is never exposed to misfortune. One falls into a feeling of security after battles, when one is drunk with success and when one believes the enemy completely disheartened. Also when a skillful enemy amuses you with pretended peace proposals. One does this through mental laziness and lack of calculation concerning the intentions of the enemy.

To proceed properly it is necessary to put oneself in his place and say: What would I do if I were the enemy? What project could I form? Make as many as possible of these projects, examine them and above all reflect on means to avert them. If you find yourself unable to cope with the situation (either because your camp is badly defended or because it is not where it should be, or that it is necessary to make a movement), put things to right at once! Often, an hour's neglect, an unfortunate delay, loses a reputation that has been acquired with a great deal of labor. Always presume that the enemy has dangerous designs and always be forehanded with the remedy. But do not let these calculations make you timid. Circumspection is good only to a certain point. A rule that I practice myself and which I have always found good is that in order to enable one to rest easy it is necessary to keep the enemy occupied. This throws them back on the defensive, and once they are placed that way they cannot again take the offensive during the entire campaign.

If you wish to be loved by your soldiers, do not lead them to slaughter. They can be spared by shortening battle by means that I shall indicate, by the skill with which you choose weakest points of attack, in not breaking your head against impracticable things which are ridiculous to attempt, in not fatiguing the soldier uselessly and in sparing him in sieges and in battles. When you seem to be most prodigal of the soldier's blood, you spare it, however, by supporting your attacks well and by pushing them with the greatest vigor to prevent time from augmenting your losses.

IX

Ruses, Strategems, Spies

RUSES are of great usefulness. They are detours which often lead more surely to the objective than the wide road which goes straight ahead. Animals have only one method of acting, but intelligent men have inexhaustible resources.

These resources are infinite. Their object is to hide your veritable design and to catch the enemy in your trap. Thus one feigns the contrary of what one wishes to do. If you open the campaign, you have your troops march and countermarch, so that the enemy cannot learn the locality where you assemble. If it is a question of capturing cities, you encamp in a place which makes him fearful for two or three of his cities at the same time. If he hastens to one flank, you throw yourself on the other. If there are no cities to be taken, but some defile you wish to seize, your ruses should tend to draw the enemy away from it giving the appearance that you are moving in some other direction. If he falls into the snare, you throw yourself at top speed on the defile you want to master.

You outwit the enemy to force him to fight, or to prevent him from it. There are two means of forcing the enemy to fight; one is in pretending to fear it. His self-confidence becomes your accomplice; security lulls him and your cunning triumphs. That is what happened before the battle of Friedberg, where roads were made for the columns from Schweidnitz to Breslau, as if the army were to retire that way at the first approach of the enemy. They thought that they only had to show themselves to chase us out. But things turned out otherwise, and they were the dupes of their preconception. The enemy is forced to fight by making marches which envelop him, which compel him to leave his camp and to move into localities where you are ready to strike him. When the Prince of Lorraine was in his strong camp at Königgrätz, he could have been forced to leave it, by his enemy making, as was possible, two marches toward Landskrom. He would have believed that his enemy had designs on Moravia, and, since he drew his food from there, he would surely have hastened toward it. Then, either while on the march or in some one of his camps, he could have been attacked.

River crossings are in this class, but I shall treat them separately.

Deceiving the Enemy

When it is desired to avoid battle, different ruses are used, and apparently offensive war is made which, nevertheless, is of opposite nature. It is your attitude that imposes on your enemy, and the suspicion that you are forming the boldest projects against him. The attitude is maintained by not withdrawing easily, and often the appearance that you are waiting for him will make him lose all desire to attack you. But if he does come you steal away by a night march, planned long before. He thinks he has you but the next day you are gone. If you only withdraw you will be followed, but then it is necessary to take a position to the flank which will prevent him from passing you without running into great danger. This sort of war is the masterpiece of the Austrians, and it is from them that it should be learned.

If you find the enemy too strong to attack, you use means to encourage him to divide his forces at the end of the campaign and in place of scattering your own men in winter quarters, you distribute them in such a fashion that in no time you can assemble the troops. Then you fall on the enemy, who are dispersed. Study the campaign that Turenne made in 1673 and study it often! It is the model of its sort.

In our times it is no longer possible to draw entire armies into ambuscades. This may have been good in other days; at present it suffices to profit from the enemy's faults. Ambuscades and other stratagems have remained useful for light troops whose fashion of fighting is favorable to this sort of ruse and who often owe their success only to the small numbers employed in them.

There is another admirable type of ruse, that of double

spies. It is necessary to direct them and then cause them to be informed of everything that you want the enemy to know. No one ever has gained greater advantage from betrayal than King William. Luxemburg had bribed the secretary of this prince; the King discovered it and made use of this traitor to give false information to Luxemburg, and this general nearly was surprised and defeated at Steenkerke.

It is essential to know what is happening among the enemy. Prince Eugene bribed the postmaster of Versailles, who opened the dispatches which went to the French army and sent him a copy. The best spies that one can have are those on the staff, or even the servants, of the enemy's general. With the Austrians, it is difficult to receive letters from their camp because of the numbers of their light troops who infest the roads.

I am of the opinion that the best thing to do against the Austrians will be to bribe some captain or major of their hussars, by means of whom intelligence can be carried on with them. Catholic priests are the best spies that one can use, but they and the common people are so accustomed to lying that they exaggerate everything, and their reports cannot be depended on. In countries where the people are opposed to you, as in Bohemia and Moravia, it is difficult to keep informed of what is happening. If greed for silver does not work, it is necessary to employ fear.

Seize some burgomaster of a city where you have a garrison, or the mayor of a village where you camp, and force him to take a disguised man, who speaks the language of the country, and under some pretext to conduct him as his servant in the enemy army. Threaten him that if he does not bring your man back, you will cut the throat of his wife and his children whom you hold under guard while waiting, and that you will have his house burned. I was obliged to employ this sad expedient in Bohemia and it succeeded for me. In general it is necessary to pay spies well and not to be miserly in that respect. A man who risks being hanged in your service merits being well paid.

X

Different Countries; Precautions

I AM writing this work only for Prussian officers. Consequently, I speak only of countries and of enemies where we may wage war. There are three sorts of countries: our own, neutral, and enemy; in enemy country there are Catholics and Protestants. All these factors have their effect in war and require different consideration according to places and to circumstances. If one takes into account thus only glory, there can be no war more favorable for acquiring it than in our own country. Since the least action of the enemy is known and discovered, detachments can be sent out boldly and, having the country on their side, are able to accomplish brilliant enterprises and are always successful; because in the larger operations of war you thus find more aid and can undertake bolder deeds, such as surprises of camps and cities. Since the people favor you, it will be due to your negligence or to an unfavorable ignorance if you do not succeed.

In neutral countries it is necessary to make friends. If you can win over the whole country so much the better. At least organize your partisans. The friendship of the neutral country is gained by requiring the soldiers to observe good discipline and by picturing your enemies as barbarous and bad intentioned. If the people are Catholic, do not speak about religion; if they are Protestant, make the people believe that a false ardor for religion attaches you to them. Use priests and the devout for this purpose. Religion becomes a dangerous arm when one knows how to make use of it. However, move more carefully with your partisans and always play a sure game.

In countries which are both enemy and Catholic, such as Bohemia and Moravia, no organization of partisans should be attempted. They will all be lost, the country being hostile to them. If you are projecting some sudden blow, it must be done by detachments. You are obliged to wage a close-fisted war and to make use of your light troops on the defensive. My own experience has convinced me of this.

XI

Kinds of Marches

THE first thing that a general must think of after having provided for the security of his camp is to have the environs and all the places through which he may need to march reconnoitered. This is done by large detachments who go to these places under other pretexts, while the quartermasters and the engineers and cavalry reconnoiter to determine in how many columns you can march. They look over the situation rapidly and make their report. The cavalry who accompany them will serve as guides to the columns on the march. The general makes his preparations on these reports.

If he is making an ordinary march forward, without fear of the enemy, this is how he disposes his command, presuming that there are routes for four columns: This evening at eight o'clock six battalions of grenadiers, a regiment of infantry, ten squadrons of dragoons, that is to say one or two regiments complete and two regiments of hussars, will form the advance guard, marching only with light baggage. The bulk of the baggage will remain with the army. They will march a mile ahead, where they will seize this defile, this height, this river, this city, or this village, and where they will wait until the army is close by to continue their march to the new camp.

The army will march tomorrow at three o'clock in four columns. The detachments return to camp as soon as the troops are to be sent into battle. Cavalry of the two right lines, marching by file to the right, form the first column; the infantry of the two right lines, marching by file to right, form the second column; the infantry of the two lines to the left, marching by file to the right, form the third column; the cavalry of the two lines to the left marching by file to the right, form the fourth column; such and such regiments of infantry of the second line and the regiment of hussars of N. will be the rear guard to cover the baggage and the artillery. These will take the two or three best roads to follow the army.

Material For Bridges

The adjutants N.N.N. remain near the baggage columns and will be responsible that the wagons are not strung out, and the officer who commands the rear guard will notify the general in time if he believes that he needs support. Three wagons loaded with beams, joists and planks to make bridges over streams march at the head of the four army columns with the detachment of carpenters. Columns should not get ahead of each other, but advance on the same front as far as is possible. The officers should observe distances exactly, and the regiments should be kept closed. If, for example, one or both the two columns of cavalry had to traverse some woods, it would be necessary to place several battalions of grenadiers at their head and even, if necessary, the cavalry could be placed in the center, assuming it is open ground and the infantry move through the woods.

If you wish to make such a march toward the enemy, you would send all the baggage under an escort to the strongest city in your vicinity. Your advance guard would only precede you by a quarter of a mile, and you would march bridle in hand, always attentive of everything that passes and always observing the ground, so that you will always have a position ready in your mind, in case you should have to occupy one in haste. That is why in these sorts of marches you should always go on the heights, as much as you are able without endangering yourself, to see the ground better and get detailed knowledge of it.

When you wish to make marches parallel to the enemy's position, you march either by the right or left in two lines, which at the same time form your two columns, and you detach a body at the extremity of the wing by which you wish to march, which serves as an advance guard. You form your rear guard just about the same, following the rear of the army. This method of marching is the surest and easiest, and especially, when it can be used, it is the best before an action because you can form in an instant. In all marches made in mountainous countries or in woods the infantry should march first. The cavalry should always go on open ground. It can act there. But in sunken roads or bushy woods you risk losing it because it is not able to fight and its weapons become useless.

Retreats with Columns Reversed

When you are marching in withdrawal before the enemy, it is essential above all to rid yourself of your heavy baggage. This done, you make your dispositions on the roads which you can use and the terrain that you occupy. If you are on open ground, the infantry withdraws first. If the country is broken, it is the cavalry that should precede. I shall presume that you have a defile behind you to pass through and that your camp contains a small plain; on this basis I would make the following disposition for two columns.

Six battalions of infantry under the orders of N. will proceed through the defile at seven o'clock this evening and post themselves on the other side, pointing their cannon and at the same time leaving the roads for the columns open. The army will march at four o'clock tomorrow morning. As soon as the troops are under arms and tents struck, the detachments, camp and village guards, etc., will withdraw and rejoin their organizations. The army will march in two columns, the cavalry of the right first, the second line of the right taking the lead, followed by the first line of the right, followed by the second line of infantry of the right, and this latter by the first line of infantry of the right, which will form one mixed column. The second line cavalry of the left wing will lead the second column, followed by the first line; the second line of infantry of the left will join the latter followed by the first line, the whole army filing by the right, six battalions of the first line sustained by ten squadrons of hussars.

This rear guard will range itself before the center of the army in two lines with intervals. When the whole army shall have left, the first line will withdraw through the intervals of the second and will place itself in battle formation not far from the defile. When it shall have established its front, the second line will withdraw, pass through the intervals of the first and pass through the defile with the larger portion of the hussars. Then the first will follow in the same manner, and the enemy, if he seeks to attack them, is contained by the troops posted the evening before to sustain the defile. These latter should remain to the end and not place themselves in march except to follow the rear guard and file through in succession. However, a general can make changes in his dispositions in accordance with the terrain and circumstances he finds.

Precautions Taken in Retreats

No attack by hussars or irregulars need be apprehended in open country. Hussars fear fire and pandours [in general, irregulars, distinguished by their guerilla tactics] hand to hand combat. In this sort of march, if they are able, they will attempt something against the baggage. They are brave when they hope to win booty. But it is quite different in woods and mountains. There, the pandours lie flat on the ground and hide themselves behind stones and trees so that they are able to fire without your being able to see where the shots come from, nor to return the injury and harm they do. Thus I shall only speak here about how you can best secure yourself from them in retreats made in mountains.

We made two such retreats during the year 1745, one from the center of Liebenthal and the other from Trautenau to Schatzlar. At that time we placed detachments of from four to six platoons on the wings of the columns. They occupied the heights which dominated the road, to turn aside these scoundrels, and the officers who were in command only had to have a few men fire against them. The rear guard also is always withdrawn from height to height following the army. But here is what happens as soon as one group abandons a height. The pandours run to it, seize it, and shoot you from there.

For this I know of no remedy, and whatever a general does he always loses lots of men uselessly in this sort of a retreat. No matter how small a plain may be, have your hussars sortie against the pandours! This turns them aside for a moment. But do not amuse yourself too much with them; otherwise your march will stretch out and you will lose prodigous numbers. Consequently, halts should not be made, and this type of difficult march should be made as rapidly as possible. But if the pandours imprudently occupy a small wood that can be turned, then hunt them down and have the hussars saber as many as you can.

If you wish to seize a post held by hussars or pandours, march against them confidently and you will be sure to carry it. Your formation is too redoubtable for them. But if they appear more determined, you may suspect that their army is nearby and able to support them. Have all sides reconnoitered, then, to inform yourself, but you may always be certain that these people should not stop you a moment, whatever attitude they assume.

What a General Can Expect

Let no one imagine that it is sufficient merely to move an army about to make the enemy regulate himself according to your movements. A general who has too presumptuous confidence in his skill runs the risk of being grossly duped. War is not an affair of chance. A great deal of knowledge, study, and meditation is necessary to conduct it well, and when blows are planned whoever contrives them with the greatest appreciation of their consequences will have a great advantage. However, to give a few rules on such a delicate matter, I should say that in general the first of two army commanders who adopts an offensive attitude almost always reduces his rival to the defensive and makes him proceed in consonance with the movements of the former.

If you were to commence the campaign first and were to make some march which indicated an extensive plan, your enemy, who would be warned to oppose it, will be obliged to adjust himself to you; but if you make a march which gives him neither suspicions nor fears, or if he should be informed that you lack the resources to execute your project, he will pay no attention to you and on his part will undertake some better considered actions which will put you in turn, in a difficult place.

Your first precaution should be to control your own subsistence. If this is well arranged, you can undertake anything. By harassing his rear with light troops, the enemy likewise can be forced to make large detachments or by making demonstrations, as if you intended to make a diversion in some other province of his realm than that in which the war is being waged. Unskillful generals race to the first trap set before them. This is why a great advantage is drawn from knowledge of your adversary, and when you know the measure of his intelligence and character you can use it to play on his weaknesses.

Everything which the enemy least expects will succeed the best. If he relies for security on a chain of mountains that he believes impracticable and you pass these by roads unknown to him, he is confused to start with, and if you press him he will not have time to recover from his consternation. In the same way, if he places himself behind a river to defend the crossing and you find some ford above or below by which to cross, this surprise will derange and confuse him. I could write much more on this topic, but what I have said should suffice and can furnish ample matter for the reflections of my readers.

XII

River Crossings

I HAVE said that the art of war is divided between force and stratagem. What cannot be done by force, must be done by stratagem. Thus if your enemy prevents you from crossing a difficult river and you are unable to cross near him, do like Caesar, Prince Eugene and Prince Charles of Lorraine. They chose a distant and easy point where they planned to cross. If the river is large, a point should be chosen where some islands confine its course by dividing it; if there are no islands, find a point where the river makes a bend. Batteries are placed there; a crossing demonstration is made at an entirely different locality to draw the enemy, and while he takes the bait, you build your bridges with all rapidity. You march there with all your forces and cross quickly.

Entrenchments are absolutely necessary in these operations. The first troops who cross should instantly have their shovels in the ground or, if woods are available, make large abatis. For greater security, a general who commands such an enterprise should create a bridgehead on both sides of the river to retain it for the time that he advances into the country to push his advantage.

Defense of Rivers

The defense of a river crossing is the worst of all assignments especially if the front that you are to defend is long; in this case defense is impracticable. To dare undertake the defense of a river crossing the front of attack should not be more than eight miles long at the most, and you should have one or two fortresses on the banks of the river at points where it is not fordable. In such a case, it is necessary to prepare against the enemy's enterprises some time in advance. I shall report the best of the expedients that have been used on such occasions.

Have all the boats and barges on the river collected and

taken to one of your fortresses, to deprive the enemy of this help. Then you personally reconnoiter the two banks of the river; that on the enemy's side to note all the places that favor his crossing; that of your own side in order to have three or more big roads made from one end to the other of your front of attack so as to be able to march conveniently and easily in several columns. And when you have found the places that you consider favorable to the enemy's crossing, if there is a small fort or a cemetery that he could use, have it destroyed and in each of these places form your troops in the order of battle and disposition you would use if you were going to fight, marking the principal points, at the same time, to be used at the right time and place.

After these precautions, camp your whole army in the center of your line of defense and make sixteen detachments, each consisting only of a few troops, and at the head of each place the elite of your officers. These detachments should be from hussars or dragoons. You will use them to scout the river bank all night. Each of these officers should be assigned a definite piece of ground for the security of which he is responsible. In the daytime it is sufficient to place mounted sentinels at the places on the banks where they are best able to observe the country.

Full, Prompt Reports

You should detach two generals from each of your wings to keep an eye on these detachments, to make certain that they will be as vigilant as such a mission requires. Officers should watch from the towers of the cities day and night to discover what is happening. Your two generals and the commanders of the cities make four reports to you a day. One should be sent at daybreak, another at ten o'clock in the morning, a third at four in the afternoon, and the last at ten o'clock at night. You put horses in relays from each of your wings to the quarters of these different generals and commanders so that in an hour and a half news from the most distant can reach you.

These detachments are only made to observe and warn.

They will inform you as soon as the enemy crosses a large body on rafts and starts to build his bridges. Your duty is to attack the enemy then. To facilitate this, you will have sent all your heavy baggage to the rear and you will be camped with one foot poised. If the enemy crosses effectively at some point, you will have your army march instantly to this point, and, since all your dispositions have been made in advance, you have only to give them to your officers at once. You should prepare yourself for infantry combat; for if the enemy is clever, he will entrench the first troops who will have crossed.

Have your dispositions well supported! Do not forget your reserves! Take advantage of the terrain and then attack brusquely and you should be able to hope for the most brilliant success. If you are required to dispute the passage of small rivers you will do it in the same fashion, only adding the precaution of destroying the fords with trees thrown into them with all their branches. If the heights are on your side, this will be a great advantage to you; if they are not, you have almost no hope of succeeding.

XIII

Surprise of Cities

CITIES can be surprised when they are badly guarded, either by detachments which are sent against them by different roads or by introducing disguised soldiers into them, as Prince Eugene did at Cremone, or after a long siege has lessened the vigilance of the governor, as did Prince Leopold of Anhalt at Glogau. Everything in war, but surprises especially, demands a great deal of information.

To make your dispositions without knowing how a city is constructed within and without, is to order a tailor to make a suit without his knowing if the man for whom he makes it is tall or short, fat or thin. Therefore, procure extensive information before undertaking anything. Cities can be surprised by detachments, as happened at Cosel. Here an officer of the garrison deserted and disclosed to the Austrians that a part of the moat was more shallow than it should have been. They entered there and found the garrison too feeble to resist them.

The surprise of Cremone was an affair planned during the winter; it would have been a brilliant opening of the campaign if it had succeeded. In our neighboring countries we actually have only two cities to surprise. One is Wittenberg in Saxony, which is a paltry fortress; the other is Olmütz, but it will not be taken, I suspect, except by formal seige. When cities are large and lightly garrisoned, it is only necessary to attack them on all sides to master them. The garrison cannot resist everywhere, and the forcing of one point carries with it the loss of all the rest. It was thus that the Emperor Charles VII, seconded by the French and Saxons, made himself master of Prague. Little cities can be taken with light ordnance. Only observe, that if you bring up cannon to destroy the gates, the cannon should be covered so that the enemy's small-arms fire cannot kill the cannoneers.

XIV

Attack and Defense; Fortified Places

THE art of conducting seiges has become a profession like those of the carpenter and shoemaker. The rules are so well known that it is not worth the trouble of repeating them. Many books tell of them, and everyone knows that a covered place is sought for the tail of the trench, that the first parallel is made as close as possible to the fort, and the work which is done in the following days is so minutely detailed and subject to such an exact calculation that it is almost possible to tell in advance the day that a fortification will be carried, at least unless this fort is singularly constructed and unless it happens to be defended by a man of distinguished merit. I shall not repeat everything, then, that Vauban and the late Prince of Anhalt have said on this subject. I shall content myself by simply adding a few ideas which have come to me, both for the attack and the defense.

I have reflected on the attack of fortified places where the terrain allows the formation of several attacks, and which also have dry moats and few advanced works, like Olmütz and Wittenberg. If you lay seige to such a place, you will open the trench on one side and make a false attack on another. As soon as the enemy shall have taken all his precautions against these two points of attack and you have confirmed him in this opinion, why should you not order an assault at an entirely different point and have all your batteries open fire at daylight, as if one of these two attacks were intended to force a way through. You use this fire as a signal to notify the detachment which, during darkness, will have approached from another side of the city to scale it.

I am convinced that little resistance will be encountered there and that the enterprise will succeed, at least if the dispositions have been based on knowledge of the place. The beseiged, thinking only of the avowed attacks, will without doubt strip all the rest of the fortifications. But to undertake such a blow you should have pressing reasons to become master of it quickly, and the beseiged must be so lulled that the idea of your design can never occur to them.

For the defense of cities nothing is stronger than mines and streams which are used to flood the trenches and fill up dry moats when the beseigers commence to make galleries in them.

Gaining Time

The art of defending fortified places consists in putting off the moment of their reduction. Thus all the science of governors and commanders of fortified cities reduces itself to gaining time by disputing the ground with the enemy. But the means that are employed are of different natures. Some officers think a great deal of sorties, and it is of them that I shall speak. I admit that I find it strange that mediocre garrisons hazard them, since a single man that they lose is more important than the loss of ten men to the beseiger. Therefore I believe that a commander should not make large sorties except when help is coming to drive away the attacker. In this case he should not spare his garrison and should regard the advantages that it gains over the beseigers like fortunate auguries of the battle which will follow. But if he is not expecting aid promptly, he can gain time in a more secure manner.

Everyone who has conducted seiges will have seen how the laborers become confused at the first shots of the enemy and how, once they have fled, the labor of the whole night is lost. This truth admitted, it follows from it that frequent small sorties, which continually alarm the beseigers, will be more successful and more sure to delay their work than large sorties, which you can never make without risking the loss of many men; and the consequence of the latter will be that you will have almost no garrison when it becomes time to defend your works. It is exactly for these works that the garrison must be conserved.

And now I shall say how I think this should be done. When you are expecting an assault against the exterior wall of the ditch, garrison it very weakly, but place your prepared fire on the works which are behind it, and in the laterals. When the assault is commenced make it hot for the enemy and, when he tries to enter, make two sorties against his two flanks; you will regain your counterscarp at once. If I were the commander, I would practice this maneuver as often as I could.

XV

Battles and Surprises

IT is very difficult to surprise encampments in our days in view of the numbers of light troops used and the precautions which are universally in practice; for, when two armies camp close to each other, either their fates are decided promptly by a battle or else one of the armies occupies a position so favorable that it can neither be forced nor surprised. Surprise, then, can happen only rarely, and more probably between detachments than between armies. However, since it is pertinent to speak of it, I shall detail what should be done in case an occasion is presented.

Your first efforts should be devoted to gaining all the information possible on the situation of the camp you wish to surprise; the second, to reconnoiter thoroughly the roads by which it can be reached; the third, to have sure and faithful guides conduct the column; the fourth, to observe inviolable secrecy and put the curious off the track; the fifth, to push all your light troops ahead under a specious pretext, but effectively to prevent some deserter from your own troops from betraying you, and to enclose the enemy in such a fashion that he cannot send his patrols out too freely; the sixth, to make good dispositions and instruct all the officers who are to execute it minutely; the seventh, to march at night without noise, with absolute prohibition against smoking or leaving ranks; the eighth, to form your line of battle not more than a quarter of a mile from the enemy camp and to attack half an hour before daylight.

If the ground permitted I should use a large body of light troops and cavalry against the camp to spread terror and confusion, and on the side where the cavalry is, I should have infantry and cannon fire to spread confusion and to prevent the squadrons from forming. However, the more that darkness is avoided the better. Daybreak is favorable because you can recognize one another, you do not risk killing your own men, and the cowards who think they can run away in the shadows, are not able to do it as well when the officers are able to distinguish them. Having carried the enemy camp, the cavalry should pursue the fugitives a certain distance. Above all, the soldiers should be prevented from pillaging and getting drunk after this success, for the enemy may be able to recover from his terror and profit from your confusion.

How to Prevent Surprises

We are accustomed, as I have said, to camp with a detached corps a certain distance ahead of the army; we also garrison all the villages up to a quarter of a mile away in front of the camp; we place cavalry guards in the rear, then infantry guards for the defense of the camp proper. Every night, besides, eight or ten patrols are made on all sides around the camp so we may be warned of what is happening.

If in spite of these precautions the enemy profits from the negligence of some subaltern officer, the troops should occupy their proper field of battle, the cavalry should charge briskly whatever it finds itself in contact, and the infantry should limit itself simply to holding its ground until daylight, when you can take measures which circumstances indicate to you as more or less favorable. In general, I believe that night attacks are only good when you are so weak that you do not dare attack the enemy in daylight. It was this reason which obliged Charles XII to attack the Prince of Anhalt at night in the affair that he had with him on the island of Rugen.

XVI

Attack on Entrenchments

IF you have resolved to attack an entrenched enemy, do it at once and do not allow him time to perfect his works. The principal point for success in your design is to know the strength and weakness of the entrenchment thoroughly. It is the general's all-around knowledge of the situation that must decide this. If you take a bull by the horns your task will be difficult and perhaps it will not succeed at all. Ordinarily, the defects of entrenchments are either that they are not solidly enough supported, or they are too extensive for the number of troops that guard them. It is upon this knowledge that you should make your dispositions. In the first case, either the enemy, instead of pushing its trench and its parapet all the way to a river, does not reach it at all, or else the river is fordable and allows you to turn him.

Two entrenchments have been taken because of this same fault; those of the French in front of Turin where the right extended to the Doire without actually resting on it. The Prince of Anhalt, who was attacking on this side, noticed it, enveloped them, and the French gave way. The other was that of the Swedes before Stralsund. The left of this entrenchment reached the sea; but Gaudi, a Prussian officer, and Köppen estimated that the water close to land was fordable on this flank. The plans were made on this information and had all the success in the world, for the Swedes were driven out. At the battle of Malplaquet the Allies lost 20,000 men unnecessarily because they chose their point of attack badly. After three hours of stubborn fighting they noticed that the abatis which covered the French left could be turned. They directed their forces in that direction, and the French were vanquished.

True and Feigned Attacks

When the entrenchments that the enemy guards are too extensive, form your true attacks and feign false ones to contain the enemy and prevent him from throwing reinforcements toward the point where you propose to make your principal effort.

The formations of your troops can be varied infinitely. I shall limit myself to proposing one for an army of fifty battalions and one hundred squadrons. I first form a line of infantry of thirty battalions, with which I compose a wing on the flank where I intend to make my principal effort. Of the twenty remaining battalions, I use twelve in the principal attack and eight in the secondary attack.

I range these twelve battalions in two lines with all my infantry three deep, and my cavalry 300 paces behind the infantry, as the following figure shows.

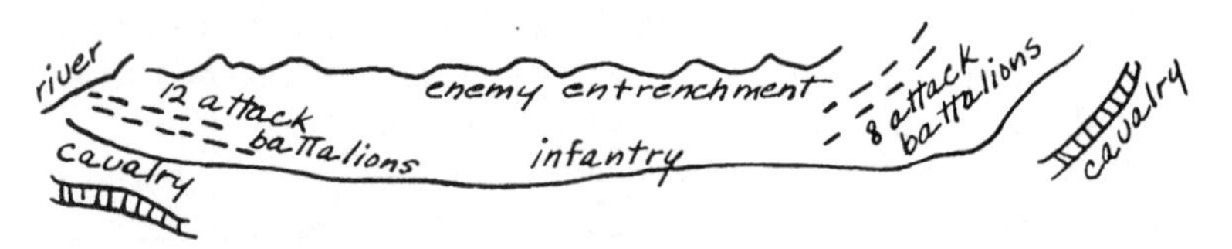

In this order of battle you can see that my line of infantry holds the enemy in check and that it is ready to profit from the least false movement that he may make. Close by the battalions which attack I order workers and men who carry fascines to fill the ditch. If the infantry captures the entrenchment, I reinforce it with all my infantry with orders to hold the parapet without pursuing the enemy. Laborers make openings so that the cavalry can enter. The latter will form themselves according to regulations, under the protection of the infantry, and will then attack as the occasion and circumstances require. (Note: The above and following sketches were made by Frederick and contained in his original manuscript. In these reproductions the inscriptions which he wrote in French have been changed to English.)

Defense of an Entrenchment

I have already said, and I repeat it, that I would never put myself in an entrenchment, at least unless a terrible misfortune, such as the loss of a battle or a triple superiority on the part of the enemy, forced me to do it. Even when inferior by half there are some resources of which I shall speak later. But supposing that very strong reasons should oblige a Prussian officer to entrench, it is essential that the front be contracted as much as possible so that the entrenchment can be garrisoned as it should be.

Furthermore, it is necessary to save two large reserves of infantry so as to be able to move them to points where they may be needed. The cavalry should be posted in a third line behind these reserves. First, attention should be given to strong support of the entrenchment, as I have said above; second, to locating it so that its flanks are strong; and third, to constructing wide, deep ditches.

If you have time, you will increase the fortifications of your entrenchments every day, either with palisades, chevaux-de-frise, concealed pits, etc. Entrenchments are a species of fortification which, in consequence, should be subordinated to the rules of this art. These consist in using the terrain well, in obliging the enemy to approach you on a narrow front, and in limiting him to points of attack which become only preliminaries in his enterprises. This is done by pushing some salients in advance which will take him in the flank if he attacks between them. By these dispositions you will reduce him to the point where he does what you want, and your mind will be distracted by fewer objects in your defense. Here is one plan of entrenchments.

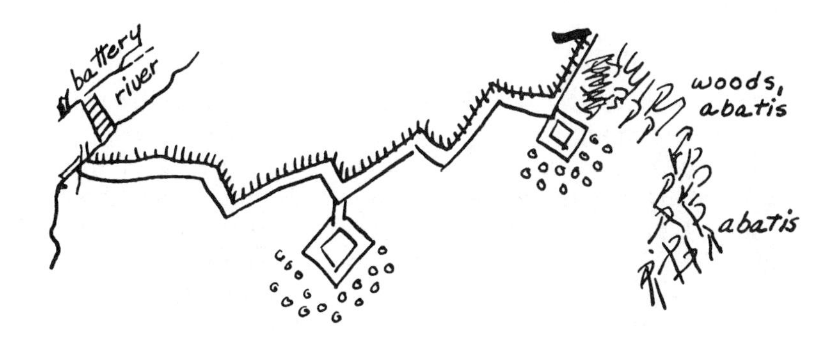

You can see that the terrain in front narrows here. The right is defended by a battery on the other side of the river which lashes the flank. The center is defended by a redoubt and the left the same, and the entrenchment, making a bend behind the abatis, prevents it from being turned. Necessarily, therefore, one of these redoubts must be attacked. Consequently, they are better fortified than the rest, and the enemy will run his nose against them.

Here is the plan of another entrenchment:

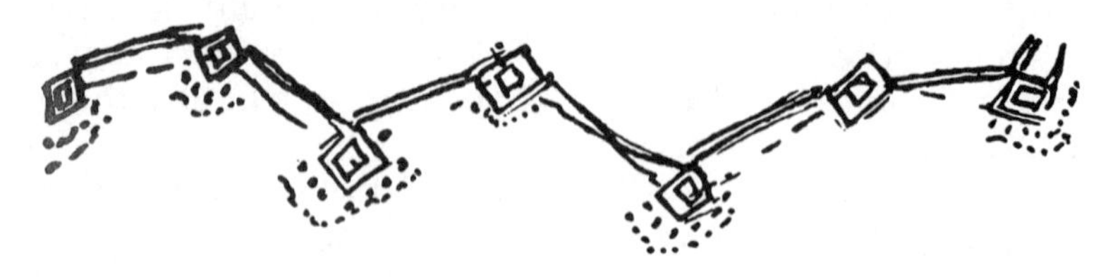

You see that some redoubts advance in salients and that they are flanked by others behind them. You see that I fortify these redoubts better than the rest of the entrenchment. Here is the reason for it: these are the points that will be attacked. I have one reason more in placing these redoubts. It is that thorough fortification of a whole entrenchment requires infinite time, and seven or eight redoubts can be perfected more quickly. Observe furthermore that the plan of the entrenchment should be according to the terrain and that your redoubts should not be more distant than 400 paces from each other so that small-arm fire is able to defend them.

Why Entrenchments are Taken

It is because whoever is enclosed in them is restricted to one ground and whoever attacks can maneuver freely; he who attacks is bolder than the one who defends himself; and because, furthermore, if a point in your entrenchment is forced, all the rest is lost on account of the discouragement that this occasions among the troops. However, I am of the opinion that the Prussians, with a resolute man at their head, could easily set right a misfortune of that type, especially if the general has conserved the resources of the reserves. Troops defending an entrenchment should fire continually, those attacking should not fire at all, but advance resolutely, gun on shoulder.

XVII

Defeating Enemy With Unequal Force

A GENERAL should choose his ground with regard to the numbers and types of his troops and the strength of the enemy. If he is the stronger and has a great deal of cavalry, he will seek the plains, primarily because his cavalry can act best there and, in the second place, because his superiority gives him the means to envelop an enemy on open ground, something he would be unable to do in broken country.

If, on the contrary, you are inferior in numbers do not despair of winning, but do not expect any other success than that gained by your skill. It is necessary to seek mountainous country and use artifices, so that if you were to be forced to battle, the enemy would not be able to face you with a front superior to your own, and so that you may be able definitely to protect your flanks.

If the terrain had not favored us in this fashion at Soor, we should never have defeated the Austrians. We were very weak in truth, but the enemy could not envelop us. His numbers, in lieu of being useful to him, became a burden. Thus the terrain equalized between us that which force had decided in favor of the Austrians.

There is the first remark which concerns only the terrain. The second relates to the manner of attacking. All weak armies attacking stronger ones should use the oblique order, and it is the best that can be employed in outpost engagements; for in setting yourself to defeat a wing and in taking a whole army in the flank, the battle is settled at the start. Cast your eyes on this plan. (See following page.)

You see how I fortify my right with which I want to make my principal effort. I have placed a body of infantry on the flank of the cavalry to fire from the woods on the enemy cavalry. I have three lines of cavalry and three lines of infantry, and my left is only for the purpose of containing the enemy's right wing, while all my forces act on the right. By this means one part of my army defeats my antagonist,

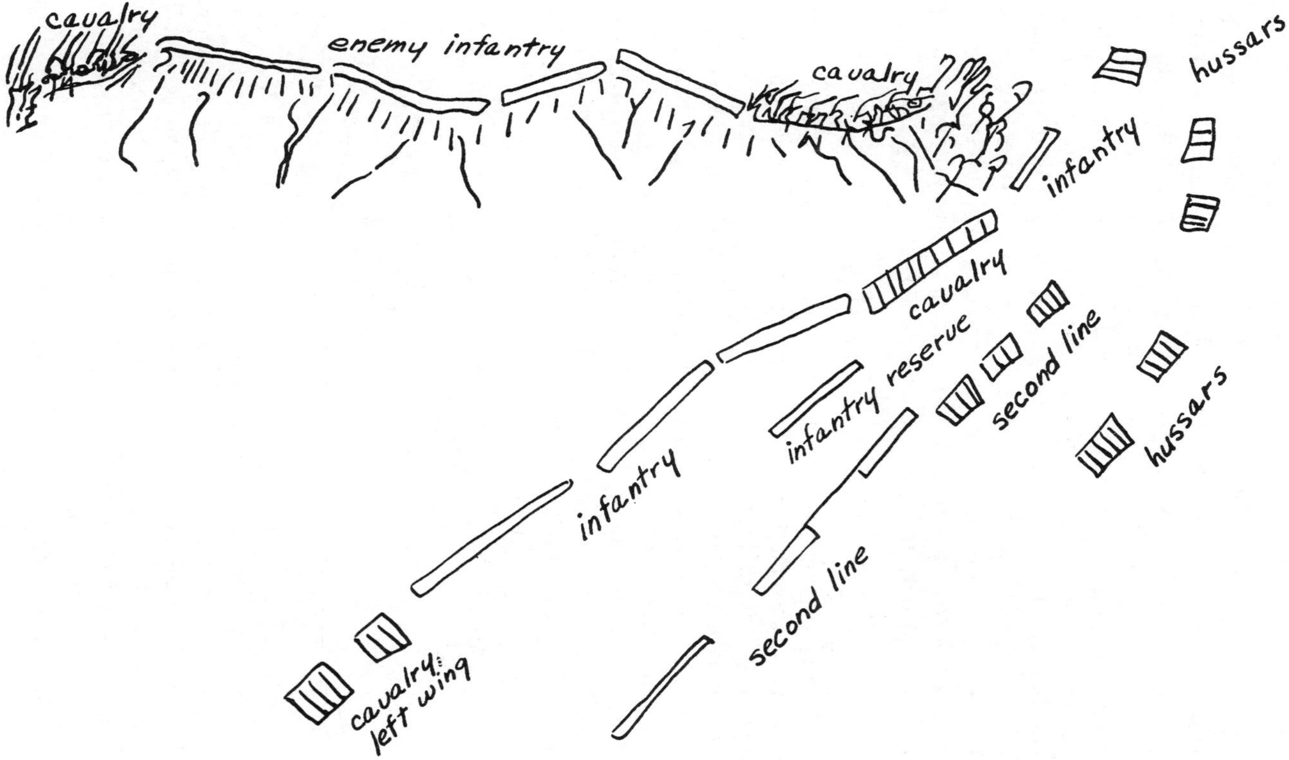

cavalry
enemy infantry
cavalry
hussars
infantry
cavalry
infantry reserve
second line
hussars
infantry
second line
cavalry
left wing

I am victorious and I execute with one part what others do with the whole. In this battle I prefer to attack with my right rather than with my left because I avoid attacking a village, which would cause too many casualties. Whenever you engage in a battle with one flank, you are the master of your army; you can speed up or slow down the combat as you deem appropriate, and the whole wing which is not fighting acts as a reserve for you. Never forget to husband all the resources you are able on every occasion and to have, in consequence, reserves always at hand to repair disorder, if it occurs at some point.

Battles in Position

Armies in position are attacked in the same manner that I have described here. It is necessary to observe that the first point is to know the locality well. However, I should advise avoiding, in so far as possible, beating against villages because this is very deadly. There are some generals who maintain that an army in position should be attacked in the center. As for me I am of the opinion that it should be attacked at the weakest point. Here is the formation of those who want to attack the center:

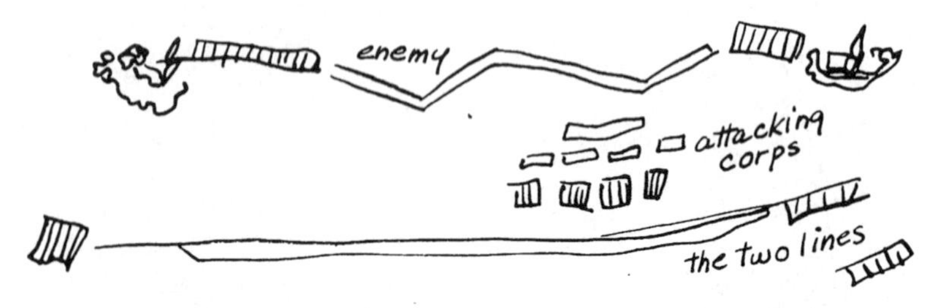

They say that in breaking the center you separate the entire army and place yourself in a situation to gain the most brilliant advantages. As for me, I repeat, I approve of all methods of attacking provided they are directed at the point where the enemy's army is weakest and where the terrain favors them the least.

Since the numerous artillery which is used these days

frequently requires the attack of batteries firing case shot, I dare venture my opinion on this subject based on the reflections which similar attacks at Soor and Kesselsdorf have inspired. Nothing is more dangerous for the infantry than to carry hostile batteries, when fourteen or sixteen cannon, advantageously placed and supported by infantry, fire several successive discharges on the assailants. This upsets all the order of our troops. Case shot makes terrible ravages; a battalion is shot full of holes before it approaches the enemy; some entire platoons are carried away and often numbers of officers. And since the power of our battalions lies in their close ranks, and each fills the place that he should hold, it is necessary to close the battalions in the act of attacking. While this is going on you are exposed to a new discharge which, causing fresh confusion, sometimes awakens natural human fears too sharply. I am, therefore, of the opinion that if the enemy holds his position, assuming that his flank cannot be turned, that he cannot be assaulted and taken.

But I noted one thing at Soor and Kesselsdorf, which convinces me that it will always be the same. This is, that in these two battles, when the losses of our men made them give way in the first attack, the enemy, wishing to pursue them, left their lines. This made their artillery hold its fire, and as a result of this our men, mingling, so to speak, with the enemy, carried the batteries at once. I desire, therefore, that the first battalion should be ordered to attack weakly to induce the enemy to quit his position, to spare the blood of the soldiers, and to gain our ends more quickly. However, if the position can be turned, this is always the surest way. Here (on the following page) is approximately the order of battle for such an attack.

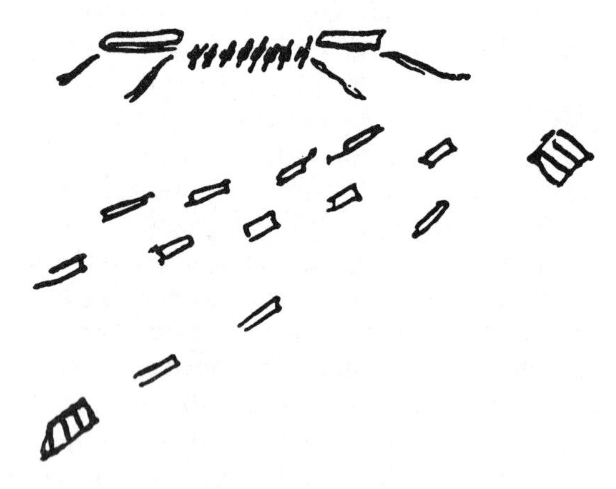

Attack. I place intervals between battalions so that they can retire through them without creating confusion in the others and I add the cavalry to charge the enemy infantry in case it leaves its position.

XVIII
Defense of Positions

I HAVE often said that for Prussians I would choose only unassailable positions, or else I would not occupy them at all, for we have too many advantages in attacking to deprive ourselves of them gratuitously. But since it is necessary to speak of this subject, I shall add only some reflections. The first is that it is essential that the wings should be so well secured that it is absolutely impossible to turn them. In the second place, that the cavalry should be in the second line; in the third place, that you conserve some reserves of infantry and fourthly, that indefensible villages should not be garrisoned, and that there is no good small, well-walled fort or a good cementery which dominates the plain and which is able to resist cannon fire, or you will lose your troops.

If, however, there are any villages on your front or on your wing, watch the wind! If it comes from the side or toward the enemy, set fire to these villages! You will save your men and make this place inaccessible to the enemy. I believe, furthermore, that if contrary to your expectations the enemy moves to attack you in such a position, you should not stick in. Use it to confound him with your cannon and then march straight at him to attack. Such a decision will embarrass him greatly and he will find himself, without knowing how, an assailant assailed. What should be watched especially is to place yourself in such a fashion that all your troops are able to act. Our position at Grottkau was worthless because only our right wing could act. The left was behind a marsh and became useless. Villeroi was defeated at Ramillies as a result of having posted himself thus.

Plains With Woods and Marshes

All battles here are basically of the same nature. Wherever the ground is broken, the country is made for artifice, and on such occasions I always favor battle for the reason that

I have given previously. Whether it is with one of your wings, or with a body intended to attack the center, it all amounts to the same thing. In our ordinary battles I have prescribed that our first line of cavalry should have a distance of only four paces between squadrons; but in country where there are canals and ditches it is necessary to give them intervals of fifteen paces, so that if some squadron encounters a ditch this will not derange the whole line. The first line of infantry never has more interval between battalions than is required to allow the cannon to fire.

XIX

Battle in Open Field

THIS sort of action should be general because you should keep your enemy occupied everywhere so that he cannot make movements that would be dangerous to you. The principal thing to watch is to block his wings. If you cannot block them both, I shall indicate the means to substitute. It is necessary, in addition, to reinforce the flanks of your infantry, so that in case one wing of cavalry gives way all your infantry will be able to support it equally. Here is my order of battle: (See following page.)

I place the infantry at the extremity of my wings to support the cavalry and to prevent the enemy from pursuing it, if it should be defeated. If it is victorious, I can use it with what I have on my flank to take the enemy infantry in the flank. I shall relate here what should be watched in the maneuver of all battles, especially in those in open field.

If you march by wings, the platoons should keep equal distances so that your whole army may be formed against the enemy by a single movement. But if you march by columns in line, the platoons must be closed on each other. Then the battalions can form, and you make your order of battle in full march, the heads of the columns drawing off to the right while the tails of the first lines deploy equally.

It is a great advantage on the day of action to be formed first. Our troops will always have it because of the skill and promptness with which they maneuver. (I am speaking here of the last order of battle.) While the army is forming, all the cannon will fire as briskly as possible. Generals who command the wings will give particular attention to supporting them well; if this is absolutely impossible for one of the wings, the general of cavalry who commands the second line will extend the first with two or three squadrons, and the hussars of the third line will extend the second with two or three. This will guarantee the flank of the cavalry in such manner that if the enemy were to fall on it they would be taken in the flank themselves.

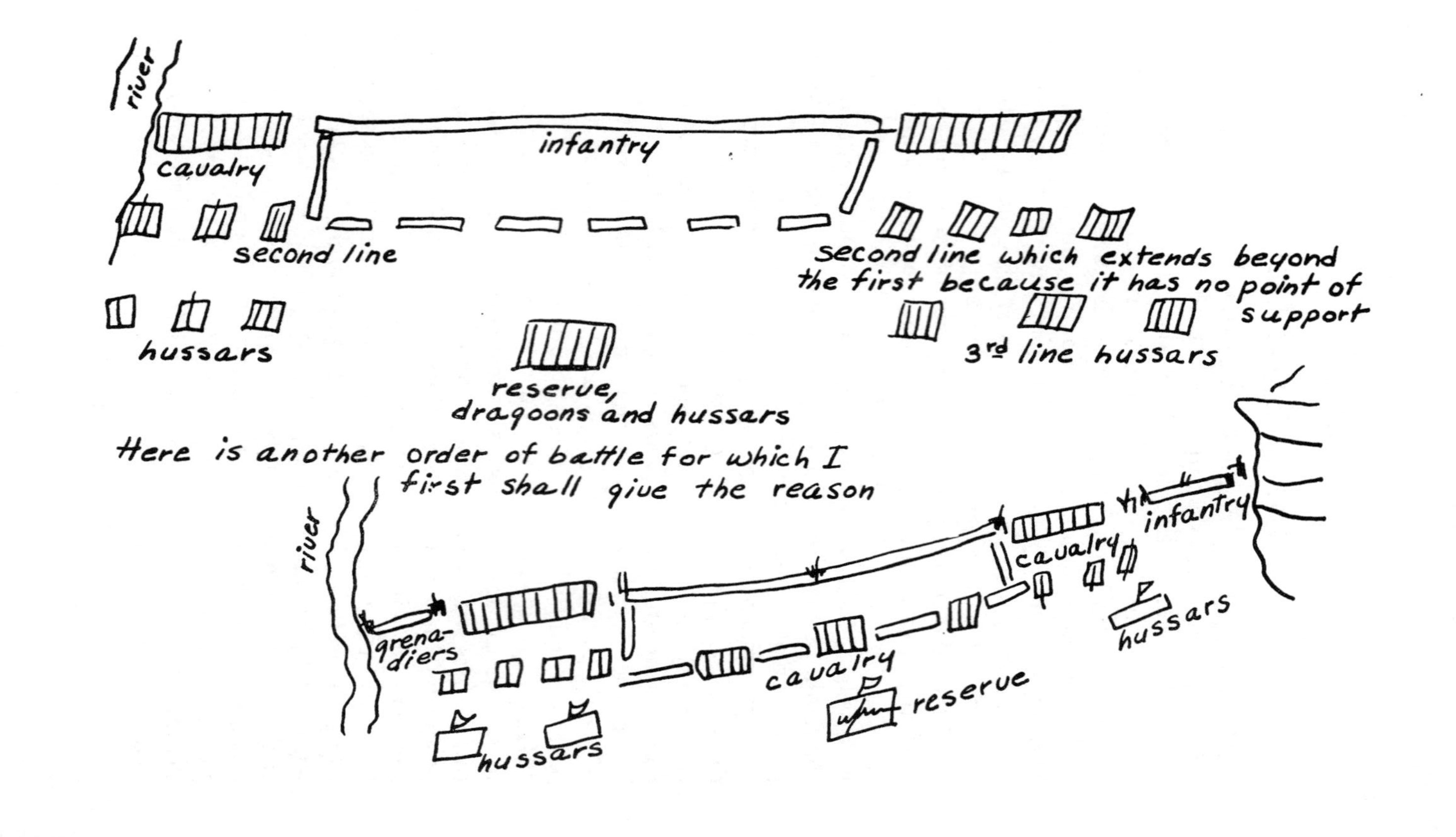
river
infantry
cavalry
second line
second line which extends beyond the first because it has no point of support
hussars
3rd line hussars
reserve,
dragoons and hussars
Here is another order of battle for which I first shall give the reason
river
infantry
cavalry
grena-diers
hussars
cavalry
reserve
hussars

Cavalry officers should always make this movement without even being ordered to. The infantry that is placed entirely on the wing of the cavalry should outflank the hostile army and fire enfilade on the hostile cavalry. I do not provide any interval for my cavalry because then each squadron would have two flanks, and a contiguous line has only one, which is its extremity. The attack of such a body thus becomes much more formidable.

Attack by Cavalry

The duty of the cavalry on a day of action is to attack, as soon as ordered, in the manner that I have prescribed in my *Institution Militaires.* It will certainly disconcert the enemy. They also should pursue when the enemy cavalry is entirely scattered.

The second line and the hussars should prepare themselves to cut the retreat of the infantry, which soon will have been scattered also. The cavalry battle will be entirely decided before the infantry can come into contact.

The general who has already gained these premonitory signs of victory can then boldly bring up the infantry which covers his wing and the infantry on his flank, to outflank the hostile infantry, take them in the flank and envelop them, if he is able. The infantry will advance rapidly, but in order. I do not want it to fire. Its menace will defeat the enemy. But in case it fires, the general should always make it continue the advance and, as soon as the enemy commences to rally around their colors, have openings made in the first line to make way for the dragoons that I have placed between the battalions of the second line, as can be seen in the second order of battle. Then all the opposing infantry will be lost. Whoever attempts to flee before these dragoons will fall in the hands of your victorious second lines of cavalry, who await them and cut them off from the defiles. Thus your success will be complete.

I have not yet spoken of the reserve. It should be commanded by a skillful general and be placed in a locality where he can see everything. He should act on his own initiative and if he sees that one of the wings are in need of help, he should conduct a reserve there without being called. But if everything goes well, the general should employ the reserve in the pursuit. That is what I have to say on the subject of general actions, and I beg my officers to read it with attention and to impress it on their minds well.

XX
Pursuit and After Battle

THE enemy is pursued to the first defile; all the harm possible is done to him, but you should not allow yourself to become so drunk with success that you become imprudent. If the enemy is thoroughly defeated, make several marches after him and you will gain a prodigious amount of territory. But always camp in accordance with regulations!

Do not neglect the principles of foresight and know that often, puffed up with success, armies have lost the fruit of their heroism through a feeling of false security. Think also of the poor wounded of the two armies. Especially have a paternal care for your own and do not be inhuman to those of the enemy. The wounded are disposed of by sending them to hospitals, and the prisoners by sending them to a neighboring fortress under a strong escort. When an army has been defeated, it is permissible to make detachments, especially when it is a question of cutting it off or of taking two or three of its cities at the same time. But it is the conditions of the time that should determine this operation, and there is no way to prescribe a general rule.

Operations of Detachments

What is practiced in armies is done on a small scale with detachments. I add that, in the one and in the other, if a little help reaches you in the action itself, it determines the turn of fortune for you. The enemy is discouraged and his excited imagination sees the help as being at least twice as strong as it really is.

Retreats of Defeated Bodies

A battle is lost less through the loss of men than by discouragement. I make my vows to Heaven that the Prussians never shall be obliged to make retreats, but, since I should touch on this matter, I shall state my opinions. For small bodies, and in large plains, the infantry should form a square protecting the remaining cavalry, thus gaining the first defile or nearest wood.

If it is a large body, I should prefer retreat in checkerboard formation by entire lines which pass through each other; a general then can save his army if he seizes the first defile instantly, so that his troops posted there protect the retreat of the others through it. If his cavalry is discouraged, let it be sent away; for the rest of this day he will not be able to bring it against the enemy again.

XXI

How and Why to Accept Battle

THE man who does things without motive or in spite of himself is either insane or a fool. War is decided only by battles, and it is not finished except by them. Thus they have to be fought, but it should be done opportunely and with all the advantages on your side. I call it opportune when it is a question of chasing an enemy out of your country; or of driving him out of a position in a locality which prevents you from penetrating into his; when you want to force him to raise a seige; or when you are unable to make seige until you have defeated him; or, when it is a question of gaining superiority for yourself for a whole campaign, and the enemy, committing an error, gives you the opportunity to take it from him. Finally, when the enemy is to receive reinforcements, you should, if possible, defeat him before their junction.

Advantages are procured in battles every time that you determine to fight, or when a battle that you have meditated upon for a long time is a consequence of the maneuvers that you have made to bring it on. The best occasions that can offer are when you cut the enemy off from his supplies and when you choose terrain favorable to the qualities of your troops and which forces the enemy to fight where you choose. After this, what is most advantageous is to profit from a poor position of the enemy to push him out of it, but especially to occupy such positions yourself as enable you to cover a great deal of the country by small movements and so located that you will never be cut off from your own supplies nor from places which you should protect. The enemy should be attacked if he places his back to a river, or if he has but a single bad defile behind him, because he risks a great deal and you risk little, even if you have bad luck.

XXII

Hazards and Misfortunes of War

WHEN a general conducts himself with all prudence, he still can suffer ill fortune; for how many things oppose his labors! Weather, harvest, his officers, the health of his troops, blunders, the death of an officer on whom he counts, discouragement of the troops, exposure of his spies, negligence of the officers who should reconnoiter the enemy and, finally, betrayal. These are the things that should be kept continually before your eyes so as to be prepared for them and prevent good fortune from blinding us.

When we wished to march from Reichenbach to Ottmachau in 1741, rains had impaired the roads so extensively that the pontoons could hardly be hauled through the mud, which, slowing up our march by two days, made our plan fall. The same day we marched a fog came up so thick that the village guards delayed us for four hours and did not join the army until nine o'clock in the morning. And there went up in smoke a plan because of a fog that could not have been foreseen.

If you form a project for an offensive campaign and the harvest fails in the country that you wish to subjugate, you easily could fail in your enterprise. If contagious illness spreads among your troops in the middle of a campaign, this will oblige you, if it goes far, to content yourself with the defensive, instead of acting with the vigor that you had proposed at the outset.

If your orders are misunderstood and some blunder occurs, accidents resulting can be decisive on days of action. I feared such an experience the day of Friedberg. Kalckstein commanded the second line. I detached him to support Dumoulin, who was on our right wing, and I sent an adjutant to Prince Charles to direct him to take command of the second line in place of Kalchstein. The adjutant misunderstood me and told Prince Charles to form the second line with troops from the first. Fortunately, I perceived this error and had time to remedy it.

If an important detachment is entrusted to a man of judgment and if this man should become sick or be killed, there is all your business hung up; for no one should imagine that sound heads are common in armies. Generals skilled in offensive tactics are rare among us; I only know a few, and, nevertheless, it is only to these that this sort of detachment can be entrusted.

Retreats Disheartening

If you have been obliged to withdraw repeatedly before the enemy, the troops get frightened. The same thing happens if you have suffered some check; even when a good opportunity presents itself to you, it must be allowed to escape as soon as you see any wavering among the troops. I have been fortunate enough never to have been in this situation with my whole army, but after the battle of Mollwitz I saw what defeated cavalry is like and how long it takes to restore their courage.

If you have spies on whom you count and the enemy discovers them, there is your compass suddenly lost, and you are obliged to conjecture and conduct yourself only on what you see with your own eyes. If one of your officers who is supposed to be scouting is negligent, he can put the whole army in danger. That is how Neipperg was surprised at Mollwitz. He had entrusted all his security to a major of hussars who was depended upon to make a patrol on the day of battle and did not do it.

Learn then to make dispositions in such fashion that the fate of your army does not depend on the good or bad conduct of a single minor officer. Especially if you have a river to guard, place so many inspectors over the officers who make the patrols that you will never be exposed by their negligence, and that more than one person shall be responsible for the occurrence.

There is still something worse to fear; this is treachery. That made us lose Cosel; that is how Prince Eugene was betrayed in 1735 by General Stein, who was a spy for France.

Finally, in considering fortuitous events and the chapter

of accidents, one can see that a general should be skillful and lucky and that no one should believe so fully in his star as to abandon himself to it blindly. If you are lucky and trust in luck alone, even your success reduces you to the defensive; if you are unlucky, you are already on the defensive.

XXIII

Cavalry and Infantry Maneuvers

YOU will have seen by what I have had occasion to delineate concerning war that promptness contributes a great deal to success in marches and even more in battles. That is why our army is drilled in such fashion that it acts faster than others. Drill is the basis of these maneuvers which enable us to form in the twinkling of an eye, and is responsible for our speed in cavalry movements. And as for cavalry attack, I have considered it necessary to make it so fast and in such close formation for more than one raeson: (1) so that this large movement will carry the coward along with the brave man; (2) so that the cavalryman will not have time to reflect; (3) so that the power of our big horses and their speed will certainly overthrow whatever tries to resist them; and (4) to deprive the simple cavalryman of any influence in the decision of such a big affair.

So long as the line is contiguous and the squadrons well closed, it is almost impossible to come to hand-to-hand combat. These squadrons are unable to become mixed, since the enemy, being more open than we are and having more intervals, is unable to resist our shock. The force of our shock is double theirs, because they have many flanks and we only have one which the general fortifies to the extent possible, and finally because the fury of our attack disconcerts them. If they fire, they will take themselves to flight; if they attack at a slow trot, they are overthrown; if they wish to come at us with the same speed with which we attack, they come in confusion and we defeat them, as it were, in detail.

As for rapid step by the infantry and attack, rifle on the shoulder, I have some very good reasons to prefer it to any other. It is not the greater or lesser number of dead that decides an action, but the ground you gain. It is not fire, but bearing which defeats the enemy. And because the decision is gained more quickly by always marching against the enemy than by amusing yourself firing, the sooner a battle is decided, the fewer men are lost. My system is based on the idea that it is up to the infantry to expel the enemy and to push him, so to speak, off the field of battle, and that it is the cavalry which crowns the action and gives it brilliance by the number of prisoners that it is their obligation to take.

XXIV
Winter Quarters

WHEN the army returns from campaign it forms a chain of winter quarters, and this always is a continuation of the campaign. The location of winter quarters cannot be determined before the issue of the campaign is settled. To make my rules easier to understand, I shall explain the disposition of our quarters from 1744 until spring, 1745.

The mountains which separate Bohemia from Silesia determined them. The quarters at Schneideberg were under the orders of Lieutenant General Truchsess. He had posts in Friedland, where he had a redoubt on the top of the mountain, and some small detached posts on the roads which run into Silesia from Schatzlar and Braunau. The positions were fortified with abatis. All the roads and trails were blocked; each small post had hussars for scouting, and the main body which Truchsess commanded was properly the reserve of the detachments.

The road and the post at Silberberg were occupied by Dessau's regiment. The Count von Glatz was defended by General Lehwaldt with a corps similar to that of Truchsess. He made abatis, blocking the roads from Bohemia, and these abatis were guarded by infantry detachments. Lehwaldt and Truchsess were within supporting distance. The Austrians could not attack the one without fearing that the other might land on their shoulder.

Troppau and Jägerndorf provided our advanced quarters in upper Silesia, and they communicated with lower Silesia through the district of Neustadt. The remainder of the troops, which did not form part of this chain, were in winter quarters by brigades so that they could be assembled more promptly and in better order.

I should add that an army should never be separated except when it is certain that the enemy will be likewise. Chains of mountains often form barriers for winter quarters;

at other times it is rivers. Neither one nor the other should ever be trusted, for mountains can be crossed wherever goats cross, and winter freezes most rivers, which then are no longer good for anything in the system of your defense. The best chain of winter quarters is composed of fortified places, such as the French and Allies have in Flanders. I note at the same time that skillful generals never occupy positions which lend themselves to many ruses, at least if they are important Troops are rested during the winter. I love to have them well nourished, but the soldier should not have any ready money. If he has a few coins in his pocket, he thinks himself too much of a great lord to follow his profession, and he deserts at the opening of the campaign. Here is how I have regulated the matter. He receives his bread free and is given more meat than ordinarily, that is to say a pound a day. The general commanding the army should see that the army is kept up to full strength. If the army is in enemy country, this country, in the nature of things, should furnish all the recruits. If this is impossible, it should give enough money to the captains so that they can raise them in the empire. It is on these occasions where an industrious general is an eagle and where a lazy one lets everything wither away.

Campaign Preparations

The general should make sure that new uniforms for the army arrive in time toward spring, that the captains provide shoes, that the wagons, cannon carriages, etc., of the army are repaired, as well as the saddles and boots of the cavalry. In a word, he should enter into all these details and himself visit the quarters to see what is going on there, if the officers are drilling the recruits, if they are working, and to animate them with kind words and reprimands to make them do their duty.

When it will soon be time to open the campaign, he should order all the governors and commanders of large cities to keep a lookout so that officers who are not sick return to their corps, and to send correct lists of those whose health prevents them from fulfilling their service immediately. It should be said to the shame of young men that debachery and laziness often make them prefer ease to glory.

Before taking the field troops are put in cantonments. From that time the brigade order of battle should be formed and the troops should be in cantonments as if they were ranged under banners, placing the cavalry of the right under a general who receives the orders, the two lines of infantry of the right under a general, the infantry of the left under another, and, finally, the last commands the cavalry of the left. This shortens the orders so that whatever you want done will be executed with more exactness.

XXV

Winter Campaigns

ONLY absolute necessity and prospect of great advantages can excuse winter operations. Ordinarily, they ruin the troops because of the sickness by which they are followed and because, remaining constantly in action, there is no time either to recruit troops nor to clothe them. An army that is employed frequently and for a long time in such a rigorous season assuredly will not stand up well. If, however, very important motives, such as we had in 1740, 1744, and 1745, require winter campaigns, the best thing that can be done is to use all possible vigor so that they may be short.

Winter campaigns are made by marching between cantonments very close together. To make this possible, all the infantry is placed in one city, as was done by the Prince of Anhalt during his Saxon expedition. On the days of marching the army assembles and marches in columns, as usual. When the enemy is near, beds are under the stars. The troops make big fires. As these fatigues are prodigious, it is necessary to cut them short; either the winter quarters of the enemy are fallen on or their army is attacked to decide things promptly.

Winter expeditions should never be undertaken in case success depends upon formal seiges. The season does not permit it. The advantage from this sort of war is that, if it is successful, the advance is prodigious. It would perhaps have cost us four ordinary campaigns to take Silesia if we had not profited by choosing the critical time. It was devoid of troops, and this gave us the means to advance and establish the theater of war on the banks of the Neisse, whereas if spring had been awaited to act we should have advanced only to Glogau.

I except reasons of this importance, but ordinarily my maxim is to leave the troops in repose during the winter and to employ this time in reorganizing the corps of the army with all assiduity, and rather to reserve for yourself the first appearance in the field the following spring.